T0288569

Setting Military Compensation to Support Recruitment, Retention, and Performance

BETH J. ASCH

Prepared for the United States Army
Approved for public release; distribution unlimited

RAND ARROYO CENTER

For more information on this publication, visit **www.rand.org/t/RR3197**

Library of Congress Cataloging-in-Publication Data is available for this publication.
ISBN: 978-1-9774-0398-8

Published by the RAND Corporation, Santa Monica, Calif.

Cover: Reese Brown/DoD; BillionPhotos.com/Adobe Stock/.

Executive Summary

Military compensation has the dual role of recompensing members for their service and of assisting the services in meeting their readiness objectives, including attracting and retaining personnel; motivating effort; inducing members to attain the ranks, positions, and jobs where they are best suited; and eventually separating personnel at the end of their career. Drawing on a large body of research, this report examines the role of military compensation as a strategic human resource tool, how well it fulfills that role, and how it could be improved. Specifically, it examines issues related to the setting of military pay, the structure of the basic pay table, the role of special and incentive pays, and the structure of the military retirement system, especially the new blended retirement system. Key recommendations include reevaluating the pay-adjustment mechanism, considering increasing performance incentives embedded in the pay table, improving the setting of special and incentive pays to increase pay flexibility; efficiency; and performance incentives; ensuring that the continuation pay under the blended retirement system is set appropriately for officers, increasing the efficiency of the retirement system by reforming the accrual charge system, and recognizing that changes to legislation to improve officer management flexibility should also consider whether and how military compensation should change.

This report is one in a series prepared specifically to synthesize several years of research about a common theme. The intent is to provide the Army's most senior leadership with an integrated view of recent years of Army-sponsored research, research that may not have achieved its full potential impact because it was presented to the Army as a series of independent research topics and findings. By looking across three to five years of research and identifying key unifying themes and recommendations, Army leadership can gain better visibility on some key issue areas and will have an additional source of information to inform key policy decisions and planning guidance. The research in this report was sponsored by the Army as well as the Office of the Secretary of Defense and draws from more than 30 years of research.

This research was conducted within RAND Arroyo Center's Personnel, Training, and Health Program. RAND Arroyo Center, part of the RAND Corporation, is a federally funded research and development center (FFRDC) sponsored by the United States Army.

RAND operates under a "Federal-Wide Assurance" (FWA00003425) and complies with the *Code of Federal Regulations for the Protection of Human Subjects Under United States Law* (45 CFR 46), also known as "the Common Rule," as well as with the implementation guidance set forth in DoD Instruction 3216.02. As applicable, this compliance includes reviews and approvals by RAND's Institutional Review Board (the Human Subjects Protection Committee) and by the U.S. Army. The views of sources utilized in this study are solely their own and do not represent the official policy or position of U.S. Department of Defense or the U.S. government.

(This page is intentionally left blank.)

Contents

(This page is intentionally left blank.)

Figures

(This page is intentionally left blank.)

Summary

This report focuses on the role of military compensation as a strategic human resource tool, how well it fulfills that role, and how it could be improved. The purpose of the compensation system is to help the services attract and manage the force to meet their warfighting and peacetime requirements in the most efficient way possible. Ideally the compensation system provides the needed flexibility in its application of incentives to recruit and retain the right numbers and quality of personnel; to sort them into the specialties, assignments, and ranks in which they could contribute the most to readiness; to motivate personnel to do their best in the interest of the service; to separate them from service gracefully when it is in the best interest of the service to do so; and to do all of this efficiently. To assess how well the compensation system meets these objectives and how it could be improved, the report summarizes and synthesizes available research, focusing on four broad areas:

1. *Setting the level of military pay and the military pay raise*: What should be the level of military pay to be competitive? What should be the annual pay raise?
2. *Structuring the military pay table*: How should military pay be structured across and within ranks to enhance talent management? How should entry pay be set? How much should pay increase when a member is promoted, and how should this increase change as a member moves up the ranks?
3. *Special and incentive pays*: How much should be put into basic pay received by all personnel versus in special and incentive pays?
4. *Deferred compensation and military retirement*: What is the purpose of the retirement system? What problems does the new Blended Retirement System solve, and what concerns are there about this new system?

While the U.S. Army cannot act alone in reforming military compensation, it can provide the impetus for change. This report aims to provide Army leadership with context and background to make well-informed compensation decisions.

Conclusions and Recommendations

The report identifies several areas where changes could improve member satisfaction with their compensation, increase efficiency in achieving retention and other readiness objectives at less cost, and increase incentives for productivity.

Key recommendations and areas for consideration are as follows:

- Ascertain if the 70th percentile of the civilian pay for civilians with similar characteristics to military personnel continues to be the right benchmark for setting the level of military pay. In recent years, military pay has exceeded the 70th percentile for both officers and enlisted, and research shows higher relative military pay increases retention and the quality of Army recruits. The quality of recruits has also increased in recent years in all the services but the Army. The Army did not increase recruit aptitude as military and civilian pay rose in recent years, possibly because it kept recruiting quality goals constant as military pay increased or possibly because the Army chose to reduce recruiting resources. The Army needs to identify why and when it should choose a different resource strategy and recruit quality objectives than the other services.

- Reevaluate the pay-adjustment mechanism; reassess the Employment Cost Index as well as other options, including the Defense Employment Cost Index with more-recent data.
- Consider increasing performance incentives embedded in the pay table, including the possibility of a time-in-grade pay table, to address compression in the officer pay structure and to increase incentives associated with promotion.
- Improve how special and incentive pays are set to increase flexibility and efficiency.
- Ensure that continuation pay, part of the new Blended Retirement System, is set at the right level.
- Consider opening another opt-in window to permit members to choose the new Blended Retirement System, given that opt-in rates were relatively low in 2018, including for the Army.
- Reform the retirement accrual system to make it different for officers and enlisted members and by service.
- Consider ways to use compensation to induce more volunteerism and greater efficiency of compensation within the Army.
- Recognize that changes to the Defense Officer Personnel Management Act to improve management flexibility should also consider whether and how military compensation should change.

Emerging Compensation Issues

Looking ahead, emerging compensation issues will call for additional research, some of which is ongoing, as well as legislative changes. Key topics include the following:

- Recently, Congress and the Army have introduced measures to increase the flexibility of military personnel management to better reward performance and to meet emerging requirements in such fields as cyber. New flexibilities were introduced in the National Defense Authorization Act for fiscal year 2019, and Congress has requested a review of the 1980 Defense Officer Personnel Management Act, which examines the current system and offers different reform proposals. At the same time, the Army's Talent Management Task Force is also considering ways to improve management flexibility, but any effort to improve retention, performance, and talent management should also consider how the current military compensation system might need to change, because military compensation is also a critical strategic human resource tool.
- The 13th Quadrennial Review of Military Compensation is considering the advisability of replacing the current time-in-service pay table with a time-in-grade pay table.
- Implementing service-specific accrual charges with separate charges for officers and enlisted personnel to eliminate inaccuracies in the current retirement accrual system.
- There is potential to use compensation to a greater extent to induce more volunteerism within the military, such as to induce volunteers for onerous or dangerous assignments.

Acknowledgments

I would like to thank Michael Linick, director of RAND Arroyo Center's Personnel, Training, and Health program, who provided the impetus and support for this report. This report drew from others I have published with colleagues, especially John Warner, professor emeritus of Clemson University, and RAND colleagues Jim Hosek and Michael Mattock. I have greatly valued our collaborations, and I know our research and analyses were better as a result of working together. I am sincerely grateful to them. I would also like to thank Barbara Bicksler and Christine DeMartini for their input to this report. The comments on an earlier draft by Michael Linick and Chaitra Hardison, associate director of RAND Arroyo Center's Personnel, Training, and Health program, were greatly appreciated. Finally, I am grateful for the input and comments from the two reviewers of this report, Paul Hogan of the Lewin Group and Sam Kleinman formerly of the Center for Naval Analyses.

(This page is intentionally left blank.)

Understanding Military Compensation

Introduction

This report focuses on the role of military compensation as a strategic human resource tool, how well it fulfills that role, and how it could be improved. The purpose of the compensation system is to help the services achieve their overall manning objective of attracting and retaining personnel in sufficient numbers and with the necessary skills and qualifications to meet grade and experience requirements. It should also help achieve three additional human resource objectives. First, compensation, in tandem with personnel policy, should be designed to provide individuals with proper incentives to work hard and seek advancement. Second, it should help sort personnel effectively to induce the proper person, rank, or job matches, including retaining and promoting the more capable personnel to the higher ranks. An implication is that individuals with low ability or effort should be induced to leave. Another implication is that "climbing" (seeking ranks for which one is unqualified) and "slumming" (the converse of climbing) should be discouraged. Third, the system should induce personnel to stay long enough to provide a return on their training and experience and to meet mid-career and senior personnel requirements but not to stay too long. At some point, the compensation and personnel systems should provide incentives for individuals to leave when it is in the services' best interest for them to do so.

Past commissions and study groups have identified additional objectives or principles of military compensation. The compensation system should be efficient and effective in both peace and war, predictable and understandable to members, fair, and sufficiently flexible to adapt to changing technology and tactics, supply and demand, and, more generally, the environment in which the military operates (Office of the Under Secretary of Defense for Personnel and Readiness, 2018).[1]

Military compensation is an emotional issue; after all, it is the most direct way the country recognizes and recompenses people for their military service. At the same time, it is a major cost element. Military pay and benefits comprise about one-third of the U.S. Department of Defense (DoD) budget, and policymakers have expressed concern that military compensation costs crowd out service-modernization efforts (Office of the Under Secretary of Defense (Comptroller), Chief Financial Officer, 2017a). Given the sensitivity of compensation issues and its significant cost, it

[1] The objectives of military compensation are stated in the 2018 Military Compensation Background Papers as well as in the reports of the fifth (Office of the Under Secretary of Defense for Personnel and Readiness, 1984), seventh (Office of the Under Secretary of Defense for Personnel and Readiness, 1992), and eleventh (Office of the Under Secretary of Defense for Personnel and Readiness, 2012) *Quadrennial Review of Military Compensation* (QRMC) reports. The Defense Advisory Committee on Military Compensation (2006) also reviewed the objectives of military compensation.

is imperative that the compensation fulfills its human resource mission as effectively, efficiently, and fairly as possible.

This report summarizes available research findings to provide U.S. Army leadership with the background needed to make well-informed compensation decisions. It considers four broad areas regarding the setting of military compensation, asks how well the objectives of compensation are being met, and how the compensation system could be improved. The four broad areas are as follows:

1. *Setting the level of military pay and the military pay raise*: What level of military pay is needed to be competitive? What should the annual pay raise be?
2. *Structuring the military pay table*: How should military pay be structured across and within ranks to enhance talent management? How should entry pay be set?
3. *Special and incentive pays*: How much of a member's compensation should be put into basic pay received by all personnel versus in special and incentive (S&I) pays?
4. *Deferred compensation and military retirement*: What is the purpose of the retirement system? What problems does the new Blended Retirement System (BRS) solve, and what are the concerns about this new system?

Although the report aims to inform Army leadership, the research results and the discussion will often take a DoD perspective because military compensation affects all of the services, not just the Army.

Contextual Background

Structure of the Force

Before addressing the four broad questions, it is useful to review the major elements of the services' forces—in this case, the Army—because these facts inform the design of the military compensation system. The Army must sustain a very large active duty force of 476,000 personnel as of the end of fiscal year (FY) 2018 (Defense Manpower Data Center, 2018). Although Army force size has varied substantially over the 45 years of the all-volunteer force (AVF), it is consistently the case that about one-third of the U.S. military's active force is in the Army. Furthermore, the Army sustains a large selected reserve force, including members of the Army National Guard.

Sustaining these forces requires recruiting large numbers of new personnel, given the typical high turnover rates in the early career and relatively junior experience mix of the force.[2] Since the start of the AVF in 1973, Army-enlisted turnover has averaged about 15 percent per year. Figure 1 shows retention profiles, by broad occupational area, for Army-enlisted personnel (upper panel) and officers (lower panel). Enlisted personnel on average remain in service six

[2] While the force today is relatively junior, it is actually more senior, with lower turnover, than was the case before the start of the AVF. For example, prior to the start of the AVF in August 1973, Army-enlisted turnover averaged 24 percent between 1970 and 1973.

years, while officers remain in service an average of 12 years. Thus, the enlisted force especially is quite junior (DoD, Office of the Deputy Assistant Secretary of Defense for Military Community and Family Policy, 2016; Office of Personnel Management, 2017).[3] There is high turnover during and after the first enlistment term, which is about three or four years, followed by higher retention rates thereafter, especially after ten years and until 20 years of service, when the members become vested in the military's defined benefit retirement system. The higher retention rates after ten years of service can be seen in Figure 1, when the retention profile is virtually flat until year 20. The large recruiting requirement is driven by an experience mix that favors junior enlisted personnel.

A noticeable feature of the curves in Figure 1 is the similarity of the retention profiles across broad occupational areas for both enlisted personnel and for officers. This similarity suggests that special and incentive pays targeted to different occupations are generally used to create similar, rather than different, retention patterns across these occupations. Furthermore, all occupational areas show a significant and similar drop off in retention after 20 years, the vesting point for the military retirement system. However, past studies and observers have argued that the optimal career length should recognize differences in the demand for youth and vigor and for experience as well as the cost of training.[4] For example, GEN (ret.) John Vessey discusses personnel in infantry whose productive lifespan is around ten to 14 years (Vessey, 2004). The retirement system's role in force management is discussed later.

---

[3] Army average years per accession were estimated using average annual continuation rates by years of service for FY 2014 to FY 2018 provided by the Defense Manpower Data Center. To compare, the average age of the Army's active component (AC) enlisted force was 27.6 years and 29.1 years, respectively, for enlisted and officers in September 2016. The average age of federal workforce was 47.4 in year 2016, and the medial age of the U.S. labor force in 2016 was 42 years.

[4] See Warner (2006) for a discussion of some of these studies.

Figure 1. Army Enlisted and Officer Retention by DoD Occupation Category

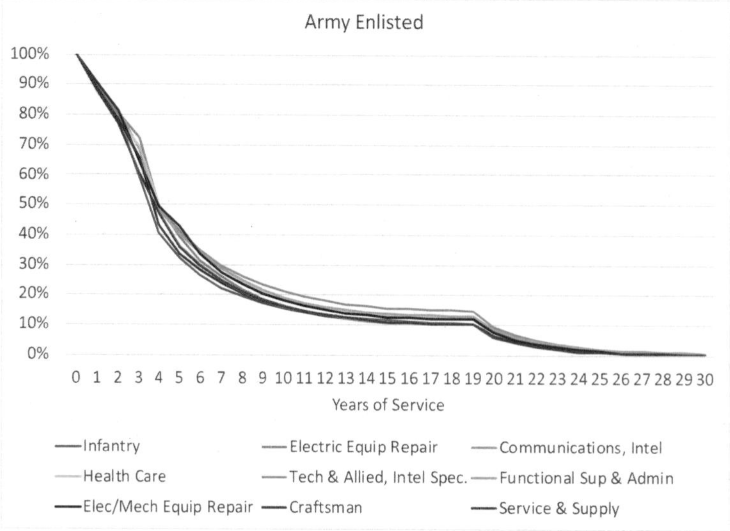

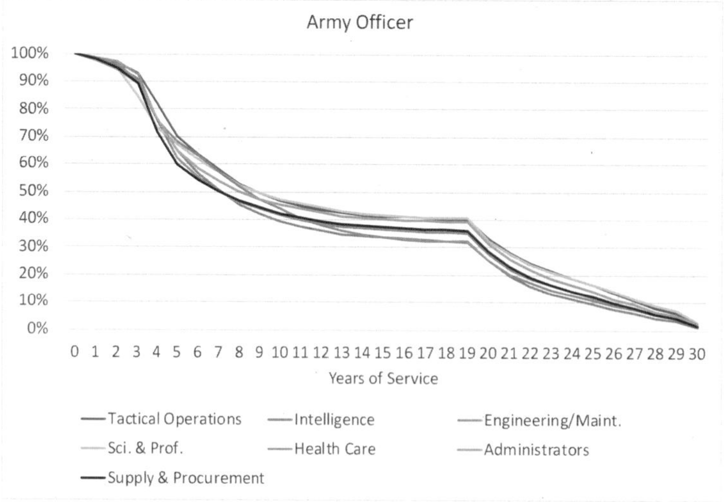

SOURCE: Author's calculation using average annual FYs 2001–2013 continuation rates for Army-enlisted personnel and officers from data provided by the Defense Manpower Data Center.

The Army has accessed about 79,000 new active-duty recruits on average over the past ten years to maintain an average enlisted force of 427,000. The Army faces substantial recruiting challenges because its recruiting goal is substantially larger than that of the other services—about double since the early 2000s. Adding to the challenge of meeting this sizable recruiting goal is that less than 30 percent of young adults are estimated to meet enlistment eligibility criteria, and, among those who do, the Army aims for high-quality entrants who have better civilian employment opportunities and prospects for college attendance (Lewin Group, 2013). A *high-quality recruit*, as defined by DoD, is an individual who scores in the top half of the Armed

Forces Qualification Test (AFQT) score distribution designated as Categories I, II, and IIIA and is a high school graduate, typically described as one with a Tier 1 education.[5]

In addition, fewer than one in ten young adults express a positive propensity to enlist in the Army (Joint Advertising and Marketing Research Studies, 2018), meaning that less than 10 percent of 16 to 21-year-olds respond "definitely" or "probably" when asked how likely is it that they would serve in active duty in the Army in the next two years. An implication of youth propensity today, or lack thereof, is that two-thirds of enlistees are recruited from the negatively propensed group.[6] Consequently, a large part of the military recruiting effort involves converting youth who are negatively propensed to join the Army into actual enlistments. Military compensation, including enlistment bonuses, together with such recruiting resources as recruiters and advertising, must be set to ensure that the Army can meet its recruiting quality and quantity requirements.

Two other stylized facts influence the design of the U.S. military compensation system. The military, including the Army, is generally a closed system with virtually no lateral entry. Vacancies are filled by young recruits without prior military service (referred to as *non-prior service*) who must be trained before they can be productive. This system is fed from the bottom, and this has an important implication. Today's officer entrants will be the generals 30 years from now. Therefore, entrants should have the ability to fill not only near-term requirements at lower levels but also future jobs at higher levels. Good generals cannot be brought in from the outside if the internal selection pool for generals is poor. The quality of the entry cohort affects the quality of future leaders. Because of the lack of lateral entry, the entry cohort should have higher abilities than the entrants in other organizations, such as the federal civil service or the private sector, because the military cannot turn to the external market to fill senior positions.

This leads to the next stylized fact: The U.S. military is a hierarchical system or pyramid, with a smaller number of personnel at each higher rank. As of the end of FY 2017, there were 1.7 majors for every lieutenant colonel, 2.2 lieutenant colonels for every colonel, and 13 colonels for every general in the Army. Such a hierarchical system means that there are fewer promotion opportunities, as personnel move up the chain of command. Because the percentage of individuals selected for promotion declines with rank, competition for promotion becomes more intense as people move up the hierarchy. From the standpoint of designing compensation, the

---

[5] The AFQT is a composite of the scores received by the applicant on the tests for arithmetic reasoning, mathematical knowledge, word knowledge, and paragraph comprehension. The AFQT score is expressed on a percentile scale that reflects the applicant's standing relative to the national population of men and women ages 18–23. *Tier 1* refers to high school graduates and nongraduates with at least 15 hours of college credit (Office of the Under Secretary of Defense for Personnel and Readiness, fiscal years 1997–2016).

[6] Using youth poll data from 2001–2003 and youth attitudinal tracking survey data from 1995–1999, Joint Advertising and Marketing Research Studies estimated that 38.8 percent of youth who are positively propensed actually enlist and 6.7 percent of the negatively propensed group enlists (Ford et al., 2009). Given that an average of 8 percent of respondents state a positive propensity to enlist (and 92 percent state a negative propensity), the implication is that 66.5 percent of enlistments come from the negatively propensed group.

rewards to promotion across the ranks should be such that the best and the brightest and those who will be future leaders are induced to remain in service and seek advancement.

Overview of the Military Compensation System

Military compensation is a complex system of pays, allowances, and benefits with components that date back to the Continental Congress. Indeed, the term *pay and allowances* was first applied to military compensation in 1789 following the adoption of the Constitution (Office of the Under Secretary of Defense for Personnel and Readiness, 2011). Basic pay is the foundation of military compensation, and every active-duty service member is entitled to basic pay, although the particular amount depends on the member's pay grade and length of service. Every member is also entitled to receive two other elements of military compensation, the basic allowance for housing (BAH, or quarters in kind) and basic allowance for subsistence (BAS, or subsistence in kind). The entitlement to these three elements—basic pay, BAH, and BAS—led the Gorham Commission in 1962 to develop the construct of regular military compensation (RMC) as a benchmark for comparing military compensation to civilian compensation. Later, the definition of RMC also included the federal tax advantage associated with receiving BAH and BAS tax-free. Basic pay is about 60 percent of RMC and is nearly half of the Army's manpower personnel budget (Figure 2). Allowances account for about 17 percent of the Army's Military Personnel Account (MPA) budget.

The services are authorized to use more than 60 S&I pays to address specific manning needs or force-management issues—although any given service member is unlikely to qualify for more than a handful of these pays. These are targeted to personnel in specific circumstances, such as those who have consistently strong civilian alternatives (e.g., medical professionals).[7] They are also offered when the demand in the military for a skill, such as language professionals, increases. S&I pay is also used when compensation is required to offset difficult aspects of the work, such as dangerous or onerous working conditions, when training costs are very high (e.g., such as for pilots), and when skills are particularly specialized (e.g., such as with chaplains). Although the number of pays is large, S&I pay represents only about 7 percent of cash compensation (Office of the Under Secretary of Defense for Personnel and Readiness, 2008) and only about 3 percent of the MPA budget (Figure 2).

---

[7] In 2008, Congress consolidated the roughly 60 categories of S&I pays into eight broad authorities to simplify and reduce the cost of administering the system of S&I pays and to give the services more flexibility to adjust S&I pays within a category. DoD was given a ten-year period to implement this policy change. The eight categories are (1) enlisted force management, (2) officer force management, (3) nuclear officer force management, (4) aviation officer force management, (5) health professional officer management, (6) hazardous duty pay, (7) assignment or special duty pay, and (8) skill incentive-proficiency pay (Office of the Undersecretary of Defense for Personnel and Readiness, 2018).

Figure 2. Distribution of FY 2019 Army MPA Across Pay Components

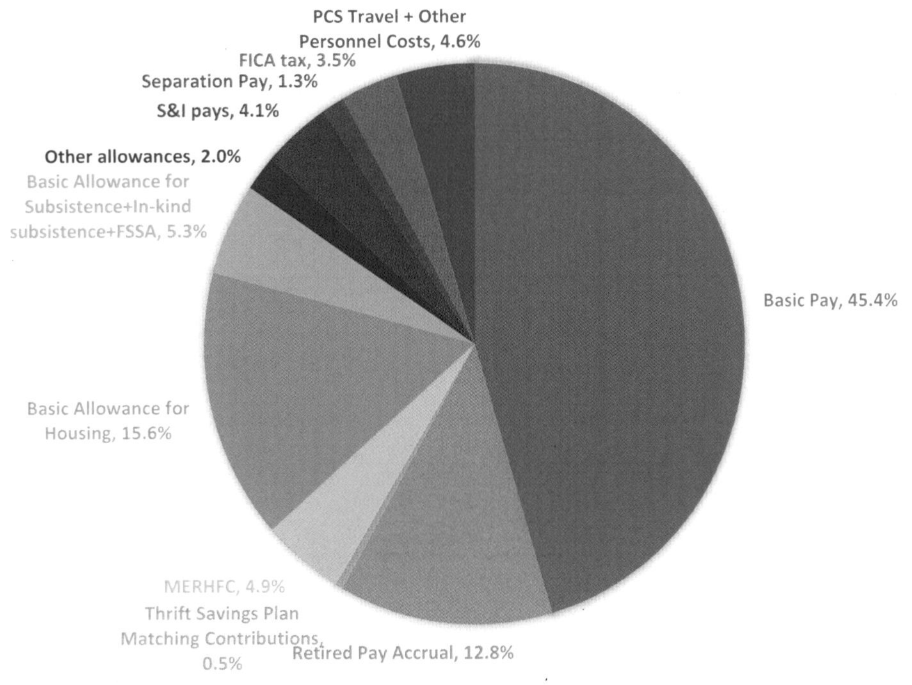

SOURCE: Author's computation using Army comptroller input (Office of the Under Secretary of Defense [Comptroller], Chief Financial Officer, 2018a).
NOTES: MERHFC = Medicare-Eligible Retiree Health Fund Contribution (made to a fund administered by the Secretary of the Treasury to finance military retiree health care programs for Medicare-eligible beneficiaries). FICA = Federal Insurance Contributions Act; PCS = permanent change of station; FSSA = Family Supplemental Subsistence Allowance.

The military retirement system is another foundational element of the compensation system, and the 20-year system dates back to the post–World War II era, when the–then U.S. Department of War sought to stem "superannuation" (e.g., a force that was too senior) following that conflict. The retirement accrual cost comprises about 14 percent of the Army's MPA budget (Figure 2). Service members become vested in the defined benefit (DB) portion after 20 years of service, which is paid out in an immediate annuity for AC members.[8] It provides a transition benefit for military retirees, as well as an old-age benefit when they exit the labor force. A new blended retirement system was implemented in 2018, which made important modifications to the 20-year system.

Other aspects of the compensation system are also noteworthy. First, military personnel receive a host of in-kind benefits, with the most major elements being health care, on-base housing, and training and education. Health care costs are particularly large—$44 billion in 2018. The military health benefit is generous compared with the health care benefits typically

---

[8] A *DB plan* is a retirement plan in which the benefit is defined by a formula, often one that uses a multiplier, final pay, and years of service. The legacy (pre-BRS) military retirement system is a DB plan with a multiplier of 2.5 percent. Under the BRS, the multiplier of the DB plan was reduced to 2 percent.

offered in the civilian sector, owing to lower premium contributions, copays, and deductibles (Office of the Under Secretary of Defense for Personnel and Readiness, 2012), leading to substantially higher utilization rate. For example, in FY 2017, TRICARE Prime (the HMO option) utilization was 57 percent, compared with 42 percent for the civilian HMO benchmark (Defense Health Agency, 2018). Past research and studies, including the 11th QRMC report, have recommended reforms to improve the efficiency of the military health benefit and thereby military compensation. A discussion of these reforms is beyond the scope of this report.

Second, despite the S&I pays targeted to specific communities, interoccupation pay differences are relatively small, largely owing to the fact that about 94 percent of cash compensation is in the form of basic pay and allowances. Third, pay varies with marital status because of higher housing allowances provided to married personnel and those with dependents and as a result of in-kind health benefits being worth more to married personnel. In 2007, the U.S. Congressional Budget Office estimated that the family premium (i.e., the percentage increase in total compensation for enlisted members with dependents), ranged from 22 percent for an E-1 to 24 percent for an E-4 to 14 percent for an E-7 (U.S. Congressional Budget Office, 2007; Hogan and Seifert, 2010).[9]

This overview shows that military personnel are paid in a variety of ways, and there are many combinations of these elements of compensation. Research, commissions, and study groups have discussed and debated how to best use the compensation system to meet manpower system quantitative targets and to satisfy the sorting, motivation, separation functions, and the other objectives of the compensation system. We turn now to consider the four broad areas listed earlier and how they might be modified to meet force staffing and readiness objectives.

Area 1: Setting the Level of Military Pay and the Military Pay Raise

Setting the Level of Military Pay

Comparability with civilian pay for the same work is sometimes the guiding principle for setting military pay. For example, pay comparability is a stated principle of military compensation in the 2011 Military Compensation Background Papers compiled by DoD (Office of the Under Secretary of Defense for Personnel and Readiness, 2011). Yet observers have noted that comparing military and civilian compensation is not easy. Jobs in the military differ from those in the private sector. Military jobs are often more hazardous and rigorous, require frequent moves, and are less flexible. Military members are subject to military discipline and are considered to be on duty at all times.

---

[9] Hogan and Seifert (2010) found evidence to suggest that the military compensation system induces military personnel to marry and divorce earlier than their civilian counterparts. Research shows that military life is disruptive to the employment and earnings of military spouses, and spouses earn lower wages than comparable civilian spouses (Hosek and Wadsworth, 2013).

Furthermore, military service provides intangible rewards such as service to the country and obtaining valuable training. Another complicating factor is the differences between the components of military and civilian compensation. Military compensation includes an array of noncash benefits (such as health care) as well a significant deferred compensation in the form of military retirement benefits. According to the U.S. Congressional Budget Office (2007), military personnel receive about 50 percent of total compensation in noncash and deferred compensation compared with about 33 percent for civilian workers.

Setting military pay for a volunteer force means that compensation should be sufficiently high to induce sufficient numbers of personnel with the required quality to voluntarily choose to serve and forego civilian opportunities. When individuals make their enlistment and retention decisions, they consider the pay of civilians with similar labor market opportunities, not just the pay of similar jobs, and those opportunities are affected by individual demographic characteristics such as age, experience, education, and gender. For this reason, pay comparisons relevant for a volunteer force are ones that account for individual and not just job characteristics.

As discussed below, comparisons of military pay with civilian pay typically use RMC as the measure of military pay. Yet basic pay is only about two-thirds of RMC, with the remainder attributable primarily to BAH and to a lesser extent BAS and the tax advantage of receiving these allowances tax free (Asch, Hosek, and Martin, 2002). BAH and BAS are not set using force-management considerations. BAH uses the results of a housing survey in local areas, while BAS is linked to the monthly cost of a food plan as defined by the U.S. Department of Agriculture. BAH varies with grade, while BAS does not. That BAH and BAS are largely out of the control of defense managers yet comprise a significant portion of RMC (as well as personnel costs, as shown in Figure 2) raises the question of whether the current pay and allowance approach to setting RMC is the most efficient approach for setting military pay. We return to this topic later in the discussion of emerging issues.

The 70th-Percentile Benchmark

Analyzing data from the 1990s, researchers found that the services were able to recruit and retain the required quality and quantity of personnel when military pay, defined as RMC, was at around the 70th percentile of civilian pay for individuals with similar demographic characteristics (see Asch and Warner, 2001). In the context of recruiting, *high quality* is defined as those with at least a high school diploma (so called Tier 1 recruits) and who score in the top half of the distribution of the AFQT. At that time, the comparison group for enlisted personnel were those individuals with a high school diploma. The 70th-percentile benchmark (rather than the average 50th percentile), reflected the unique nature of military duty discussed previously.

In its 2002 report, the ninth QRMC, reviewed the idea of a 70th-percentile benchmark and found it remained relevant but also concluded that comparing enlisted personnel with civilians with high school diplomas no longer reflected the education level of the force (Office of the Under Secretary of Defense for Personnel and Readiness, 2002). The ninth QRMC pointed out that a large percentage of the enlisted force had some college education, and the military actively recruited from the college-bound youth market. Thus, the ninth QRMC argued that analysis of pay comparability should compare military pay for enlisted personnel to the 70th percentile of pay of comparable civilians with some college because the pay of civilians with some college is the relevant civilian opportunity for most enlisted personnel. Similarly, the comparison groups for officers should be civilians with a bachelor's degree or higher.

Thus, the notion of the 70th percentile as a focal point for military pay comparisons was born, and it continues to be used as a guiding factor in setting military pay today. What is important to understand, however, is that the 70th percentile is not a goal in and of itself. Compensation should be set high enough to attract and retain the quantity and quality of personnel the services require, and the level of compensation necessary to do this may or may not be at the 70th percentile.

Using 2009 data, dating a decade after those which were used in the ninth QRMC, the 11th QRMC found that military pay exceeded the 70th percentile. Specifically, it found that RMC was at about the 90th percentile for enlisted members and at the 83rd percentile for officers.[10] Thus, over the course of the 2000s, military pay relative to civilian pay increased substantially. We discuss the details of this increase.

Military pay continues to exceed the 70th percentile, and, in fact, the percentiles for 2016 were virtually the same as what the 11th QRMC found for 2009 (Hosek et al., 2018). Given that military pay is above the 70th percentile benchmark and has been for some time, the important question is whether this benchmark is still relevant or whether military pay is set too high relative to civilian pay. Answering this question requires understanding whether the factors that led to the setting of the 70th percentile as the benchmark are still relevant today and specifically if the recruiting and retention outcomes produced today as a result of the level of military pay relative to civilian pay are the desired outcomes.

Evidence About the Relationship Between Military Pay and Recruiting and Retention Outcomes

A large number of studies using varied methods and data has estimated the responsiveness of high-quality enlistments to changes in military pay relative to civilian pay. In general terms, these estimates indicate that, as military pay increases relative to civilian pay, enlistment of high-quality youth increases. For example, using FYs 2000–2008 data, Asch et al. (2010) estimated that if the level of military pay increased by 10 percent relative to civilian wage opportunities, Army enlistment of high-quality youth would increase by 11.5 percent.

That said, a recent RAND study of the relationship between relative military pay and recruit quality for all services using data between 2001 and 2016 found a striking result for the Army;

[10] The 11th QRMC chose a mix of civilian comparison groups for enlisted personnel and officers. For enlisted, it used civilians with a high school diploma for junior enlisted, those with some college for mid-career enlisted, and those with an associate's degree for more senior enlisted. For officers, the comparison groups were civilians with four-year college degrees and those with a master's degree or higher.

the other services increased recruit quality when relative military pay rose, but this was not the case for the Army (Hosek et al., 2018). Specifically, the percentage of Army recruits in AFQT Categories I–IIIA, meaning the top half of the AFQT score distribution, tended to stay at or somewhat above the DoD benchmark of 60 percent after 2008, while the percentage greatly exceed the benchmark for the other services (Figure 3). The Army increased its Tier 1 recruiting rate after 2008 and became more selective for the non–Tier 1 recruits, bringing in those in higher AFQT categories.

The reason why the Army did not increase recruit aptitude as military pay rose relative to civilian pay is an open question. One possibility is that the introduction of the post-9/11 G.I. Bill in 2009 eliminated the Army's ability to provide education benefit "kickers" to recruits entering selected occupations; these kickers gave the Army a recruiting advantage over the other services. Without this advantage, it may have been more difficult for the Army to expand high-quality recruiting relative to the other services because recruits in all services had access to the post-9/11 G.I. Bill. Another possibility is that the Army set its recruiting-quality goals constant as military pay increased, allowing the Army to hold down such recruiting resource costs as bonuses and advertising, yet still meet the DoD benchmark. The other services focused on increasing recruit quality above the benchmarks when military compensation increased.

The different strategies adopted by the services raise the question of what level of recruit quality is appropriate and what is the proper Army response for setting recruit quality when military pay increases relative to civilian pay: Should recruit quality increase, or should recruiting resources and associated costs decrease when recruiting conditions improve as a result of increased military pay? Put differently, how should the Army reap the benefits of a less-challenging recruiting market? On one hand, the Army's approach would be more appropriate if DoD benchmarks for recruit quality are still relevant. On the other hand, if the benchmarks for recruit quality are too low and higher quality is required, the other services' approach would be more appropriate. That said, research shows that increasing resources other than military pay, including enlistment bonuses, may be a more cost-effective means of increasing recruit quality (Asch et al., 2010).

Figure 3. AFQT Category I–IIIA Enlisted Accessions by Service

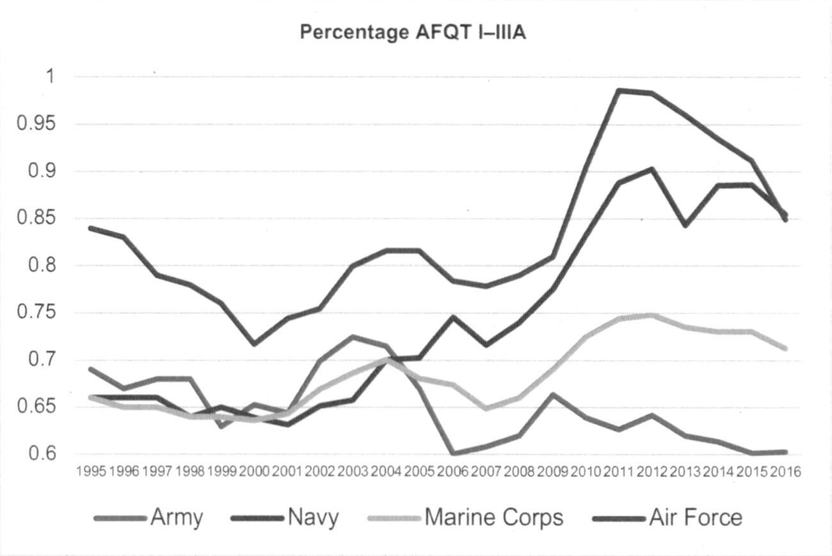

SOURCE: Office of the Under Secretary of Defense for Personnel and Readiness, fiscal years 1997–2016.

The level of military pay relative to external civilian pay has been found to be an important determinant of retention as well, although past studies have taken different empirical approaches to estimating the determinants of the retention decisions. Warner and Asch (1995) summarize pay estimates from early studies and found that if all elements of future military compensation were to rise by 10 percent and civilian compensation remained unchanged, first-term retention would be predicted to rise by about 20 percent. The increases in predicted retention rates were somewhat smaller for soldiers in their second term, as would be expected because base retention rates are relatively high beyond that point.

More-recent studies use data that track the individual retention decisions over the military career of military entrants. This approach, known as the dynamic retention model (DRM) as first posited by Gotz and McCall (1984), is advantageous because it uses a structural model that provides estimates of the causal parameters underlying retention decisions. Estimating causal parameters is especially important when the compensation policies being analyzed have never been tried before. Because this approach estimates the parameters underlying the retention decisionmaking process, one can then use these estimates to perform "what if" simulations of new policy alternatives. Simulations from estimated DRM models for Army-enlisted personnel and officers show that Army retention is responsive to changes in military compensation (see, for example, Asch et al., 2008; Mattock et al., 2014; Mattock, Asch, and Hosek, 2014).

In the 2000s, Army deployments during operations in Iraq and Afghanistan were extensive, and research shows that deployments also affect retention (Hosek and Martorell, 2009). Deployment had a positive but decreasing effect on Army first- and second-term reenlistment from 1996 to 2005, but the effect turned sharply negative in 2006 because so many soldiers had been deployed for long periods of time, and those deployed longer are less likely to reenlist. Thus, extensive deployments eventually reduced reenlistment in the Army. More important, increases in military pay and bonuses (discussed below) offset reduced retention during this period. These increases were critical because they helped the Army sustain retention overall despite the stress on the force associated with extensive deployments.

Past research has also examined whether higher-aptitude personnel are more likely to leave the military (Asch, Romley, and Totten, 2005). Rigorous studies found that personnel with higher aptitude scores perform better in the military (Orvis, Childress, and Polich, 1992; "Project A: The U.S. Army Selection and Classification Project—Special Issue," 1990; Winkler, Fernandez, and Polich, 1992; Scribner et al., 1986). These personnel have better external opportunities, but they may also have better internal promotion opportunities. Using the AFQT score as the metric of personnel quality, research shows that the effects of better external opportunities are offset by the better internal opportunities; the quality of those who stay is not much different from the quality of those who leave. This implies that setting pay and other compensation and recruiting resources to recruit a high-aptitude applicant pool is especially important because the AFQT scores of the pool that is recruited drive the scores of those who are retained over their career. It also implies that military pay levels have been sufficiently high historically to retain these high-quality personnel over a career.

Is Military Pay Too High?

Given that military pay exceeds the 70th percentile, is military pay too high? Figure 3 shows that recruit quality measured in terms of AFQT exceeded the 60-percent DoD benchmark and increased after 2008 for all services except the Army. If the complexity and highly technical nature of today's defense environment requires a higher-quality force and, therefore, higher scores on the AFQT, then perhaps the 60-percent DoD benchmark for recruit aptitude is too low, implying that the 70th percentile for military pay relative to civilian pay is also not the right standard today. But, if that is the case, why did scores not increase for the Army as well?

Furthermore, military base pay is not an efficient means of increasing recruit quality to meet increased standards. It is a blunt and costly instrument for addressing recruiting challenges. Unlike investments in bonuses, recruiters, and other recruiting resources that can be targeted to particular groups of personnel or locations, increases in military pay affect all military personnel and the personnel budget of every service. Thus, if higher recruit quality is needed, it would be appropriate to reduce military pay (or slow its growth) and increase recruiting resources and special and incentive pays. In the case of the Army, if military pay were to decrease, the Army too would need to compensate by increasing recruiting resources. In the Army's case, this would be necessary to prevent recruit quality from falling below the DoD quality guidance because Army recruit quality is already at around the levels set as goals by DoD.

Setting the Annual Military Pay Raise

Related to the issue of setting the level of pay is setting the annual percentage increase. The annual pay adjustment is an important source of income growth for military members given that basic pay accounts for about two-thirds of RMC. Section 1009c of Title 37 (U.S. Code, 2003) provides a formula for the annual increase in basic pay that is indexed to the annual increase in the Employment Cost Index (ECI) for the wage and salary of private industry workers. The military pay increase is measured as the 12-month percentage change in the ECI for the third quarter of the calendar year where previous years' values are used. For example, the ECI guiding the 2020 pay raise is the percentage change for the third quarter of 2018 relative to the third quarter of 2017. That said, the statute also allows the President to specify an alternative pay adjustment so that the ECI is ultimately only a guide for adjusting basic pay. Recruiting and retention outcomes also inform the recommended pay raise.

Historically, the military pay increase has often deviated from ECI increase. The surge in the civilian economy in the 1990s because of the dot-com boom put significant strain on recruiting and retention in the late 1990s. In 1999, for example, the Army and U.S. Navy both missed their recruiting goals. In response, the National Defense Authorization Act of 2000 increased basic pay by 4.8 percent, restructured the basic pay table by providing higher pay increases in certain years of service and grades, and authorized increasing basic pay by half a percentage point above the ECI through 2006.

The higher-than-ECI increases in basic pay continued through 2010 and helped to sustain recruiting and retention during wartime. Frequent and long deployments to Iraq and Afghanistan stressed recruiting and retention, especially between 2005 and 2008, but increases in basic pay, the housing allowance, and enlistment and reenlistment bonuses enabled the Army and the other services to meet their personnel needs. The basic pay increase reverted to the percentage increase in the ECI in 2011 through 2013, was below the ECI percentage increase in 2014 through 2016, equaled the ECI percentage increase in 2017 and 2018, and exceeded the ECI percentage increase in 2019. The slower-than-ECI increases between 2011 and 2013 reflected numerous factors, including defense downsizing, a weak civilian economy because of the Great Recession, and sequestration.

Figure 4 shows the cumulative percentage changes since 2000 in the ECI, basic pay, and RMC of an E-4 and in the RMC of an O-4 averaged across grade in the case of RMC. Note that the percentage change in the ECI in each year is the percentage relevant for that year's basic pay increase. Thus, the 3.2-percent change in ECI in 2004 informed the 3.7-percent change in basic pay in 2004, although the ECI computation uses the 12-month percentage change in the third quarter of 2002 and 2003.

Figure 4. Cumulative Percentage Changes Since 2000 in ECI, Basic Pay, and RMC

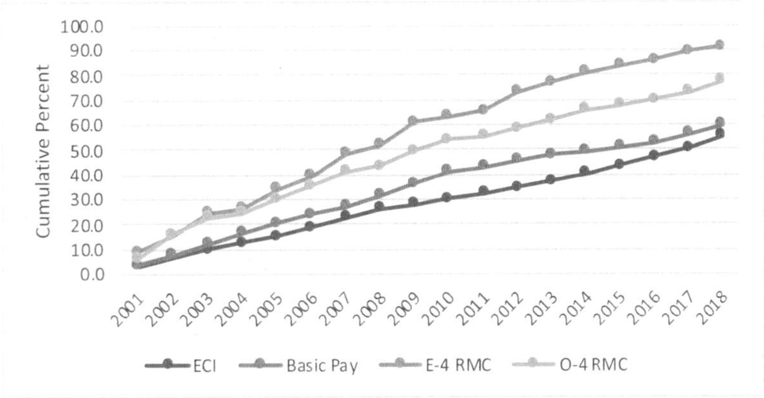

SOURCES: U.S. Department of Labor, Bureau of Labor Statistics, undated; Kapp, 2018; DoD, Office of the Under Secretary of Defense for Personnel and Readiness, Directorate of Compensation, 2000–2016.
NOTES: Basic pay increases shown are the statutory increases and do not incorporate additional increases that resulted from restructuring the military pay table in selected years (see Kapp, 2018). Percentage increases in average RMC for E-4 and for O-4.

Overall, military pay, measured as either basic pay or RMC, has grown faster than the ECI. Through 2013, after which the growth in military pay slowed relative to the growth in ECI, basic pay had grown 59.8 percent since 2000, compared with 55.3 percent in the ECI. RMC grew even faster because of the growth in the housing allowance as a result of reductions in the 2000s in the required out-of-pocket expenses for housing when computing the housing allowances. The slowdown in the growth in military pay between 2014 and 2016 reduced the growth gap between basic pay and the ECI, but the growth in basic pay was still greater since 2000.

Using the ECI to compare military with civilian pay growth has advantages because the ECI is readily available from the Bureau of Labor Statistics, and its use is codified in law. Furthermore, the ECI provides easily understandable information to military personnel about U.S. commitment to maintaining comparable pay. But to function well, the ECI as an adjustment mechanism should be accurate in the sense that it is relevant to military enlistment and reenlistment decisions. A 1992 study analyzed data between 1982 and 1991 and found that although military pay lagged substantially behind the growth in the ECI, the services recorded no major recruiting and retention problems during this period (Hosek et al., 1992). That is, measuring the "pay gap" using the ECI did not perform well in terms of tracking force-management outcomes.

The problem with the ECI is that it does not track well with the opportunity wages that are relevant to military personnel. The ECI uses data of private-industry wage and salary workers who are older and have a different mix of education and occupations than military personnel. Other pay-adjustment mechanisms that better reflect the demographics of military personnel could be used. Hosek et al. (1992) developed the Defense Employment Cost Index (DECI) for this purpose, and analysis of pay-gap comparisons using the DECI found that the DECI tracked military enlistment and reenlistment more accurately than the ECI. This makes sense because the DECI is more focused on providing a representation of the opportunity wage of military personnel. That said, the ECI is available quarterly, while the DECI is an annual metric. The ECI is also less influenced by the more variable wages of young workers than is DECI, thus it is more stable over time.

When it was proposed in 1992, the DECI did not gain acceptance from DoD or Congress, and it was never adopted. But the ECI has clear disadvantages. The functioning of the ECI as an appropriate pay-adjustment mechanism, and alternative approaches, including the DECI, should be reassessed with more-recent data.

Area 2: Structuring the Military Pay Table

The Army needs to do more than meet overall requirements. It needs the right people and the right quality at each rank and at each seniority within the ranks, and it needs the pay structure to motivate and support continued work effort. How much should pay increase when a member is promoted, and how should this increase change as a member moves up the ranks? How much should pay increase within a grade, and should the step increases vary across ranks? Put differently, how should the pay table be structured across grades (inter-rank pay spreads) and across years of service within a grade (intra-rank pay spreads)? Answers to these questions are closely tied to the importance of talent management and the objectives of motivating hard work and sorting personnel effectively (discussed in the introduction of this report).

Military personnel make choices not only about entry and retention but also about how hard to work. Although the military expends many resources monitoring work effort, it still cannot monitor effort perfectly or without cost. Pay and personnel policies can motivate personnel to work harder and more effectively only to the extent that future outcomes are made contingent on current performance, and members are responsive to such contingent compensation.

Rewards Associated with Promotion

An important source of contingent compensation are the rewards associated with promotion. Promotion to a higher rank provides both monetary and nonmonetary rewards, and monetary rewards can come either through the active-duty pay associated with higher rank or in the form of retirement or other deferred benefits (Lazear and Rosen, 1981). Another potential source of contingent compensation is pay within a grade (e.g., an intra-grade payoff that is contingent on effort). Performance bonuses or other nonmonetary rewards to top performers could spur effort. But today, intra-grade performance incentives are weakened by the lock-step nature of longevity increases in the current pay table.

The military's hierarchical rank structure affects promotions across the ranks. Subject to individual qualifications, personnel are promoted through the lower ranks with virtual certainty using time-in-grade or time-in-service requirements. But beyond the junior ranks, promotions are competitive, and only a fraction of those seeking advancement are promoted. The competition at the upper ranks is tighter because those who are eligible for promotion are increasingly similar in terms of their qualifications.[11]

By affecting promotion, the hierarchical structure also affects effort incentives. A declining probability of promotion at higher grades will tend to diminish work effort, if all else is the same, because the reward associated with promotion is affected by the likelihood that the reward will be received. If the likelihood of promotion is low, individuals will not expend much effort to be promoted without a sufficient reward for promotion.

Numerous factors affect promotion chances and in turn the incentive to work harder. Random factors such as "noise" or "luck" play a role. Because promotion in the lower ranks uses explicit criteria or standards, luck has only a small influence on promotion outcomes, so individuals are likely to work harder. As service members progress through the upper ranks, however, the role of luck increases and thereby discourages effort; having the "right" assignment or working for the "right" mentor, for example, loom larger in the promotion outcomes. The composition of the promotion pool also affects the likelihood of promotion. When the promotion pool is diverse, as it is in the lower ranks, it is easy to bypass colleagues by working harder. In the upper ranks, the reverse is true because the promotion pool becomes more similar.

Tastes for military service versus the civilian sector influence effort incentives. Individuals with a high taste for military service are more likely to stay for future periods. Consequently, they are more likely to reap the future benefits of harder work today and will therefore work harder. An important policy implication follows. Because first-termers have lower tastes for military service than careerists on average, a pay raise targeted at the first-term force will not produce as much extra effort as a raise targeted at the career force. This result provides some rationale for skewing the pay table by longevity as well as by rank, as discussed below.

Finally, up-or-out rules can also induce effort by lowering the expected payoff to remaining in a lower grade (relative to advancement to a higher rank). Up-or-out rules therefore can serve as a substitute for a direct increase in pay associated with promotion.

Sequencing Pay by Grade and Longevity

Given the relationship between effort and promotion opportunities, how should pay be sequenced by grade and longevity? The military's objective is to sharpen the competition for promotion so that the most qualified service members rise to the top. One way to accomplish this is through bigger pay spreads associated with promotion (e.g., larger intergrade pay spreads), among other policies. Such spreads motivate harder work in the quest for advancement. More important, larger spreads encourage those who are more able to remain in service and therefore help maintain the quality of the promotion pool. And by improving the talent pool and by

---

[11] Evidence on the greater similarity and higher quality of those in the upper ranks is shown in Asch and Warner (2001) and Asch, Romley, and Totten (2005).

inducing those who are more able to work harder, larger intergrade spreads prevent less-qualified members from being promoted (referred to as "climbing").

As discussed previously, as individuals progress toward senior ranks, promotion rates fall. Absent any change in the structure of pay, declining promotion rates tend to discourage effort. Clearly, the increase in pay associated with promotion needs to rise with rank to maintain effort. The tendency to reduce effort at higher grades is accentuated by several other factors. Two mentioned previously are the rising relative importance of "luck" in promotion outcomes and the increasing similarity in skills and qualification of individuals in the promotion pool. Another is that as personnel progress through the ranks, the number of remaining promotions (and therefore promotion payoffs to greater effort) that can be earned falls, so effort incentives fall as people reach the end of their career and their remaining promotion prospects.

Thus, offering significantly higher pay at higher ranks (or skewing pay), as is illustrated for enlisted pay in Figure 5,[12] is required for personnel to see a continuing reward to effort. The figure shows enlisted basic pay at promotion to each grade relative to pay at promotion to E-5. Thus, pay at promotion to E-5 is normalized to 100 percent. Pay at promotion to E-4 is 83 percent, a difference of 17 percentage points relative to E-5, while pay at promotion to E-6 is 125 percent relative to E-5, a difference of 25 percentage points relative to E-5. The spread increases with rank. Pay at promotion to E-9 is 245 percent relative to E-5, a spread of 145 percent. As shown in Figure 5, a skewed pay structure is one that is also convex.

Figure 5. Hypothetical Example: Basic Pay at Promotion Relative to Pay at Promotion of an E-5 (in Percentage)

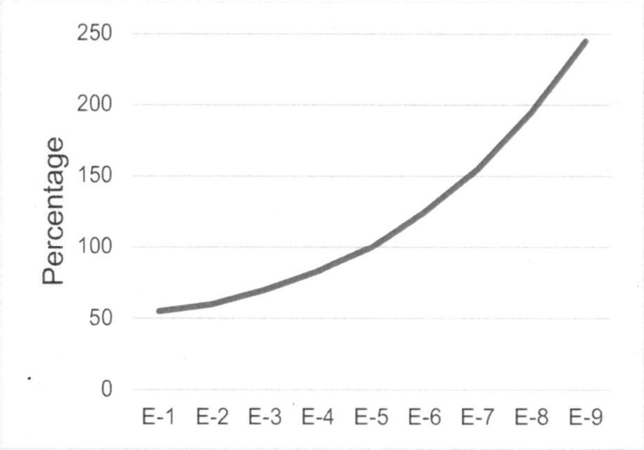

SOURCE: Author's computations using 2018 basic pay table.

---

[12] For enlisted personnel, the profile shown in Figure 5 assumes promotions at the following years of service (YOS): E-1 at one YOS, E-2 at one YOS, E-3 at two YOS, E-4 at three YOS, E-5 at eight YOS, E-6 at 13 YOS, E-7 at 20 YOS, and E-9 at 25 YOS. These promotion points use data from the 1990s (allowing the comparisons shown later in Figure 6). Faster or slower promotion shifts the curve down or up but by and large does not change the shape of the curve.

Other factors, however, reduce the required skewness. The more that individuals value the status and other nonmonetary factors associated with higher ranks, the less skewness required to motivate effort in the lower ranks. A second factor is the transferability of training. The less transferable military training is to the external market, the less skewness is needed to efficiently maintain a given level of retention. The third factor is the correlation between tastes for service and ability. If the correlation is positive so that the personnel who have stronger tastes for military service are also the more able, then less skewness is required to induce the more able to stay and seek the higher ranks.

Another factor is team-oriented work. Rosen (1992) writes that

> If rewards are skewed too much, competitors may take steps to make others look bad rather than making themselves look good. Lack of cooperation and reduced cohesiveness can reduce the effectiveness of the overall team. . . .Some happy medium must be struck here.

The argument about making others look bad is not as relevant in the military case because of the sheer numbers of individuals participating in the promotion contests and their geographic dispersion (sabotage is more likely when people work together). However, insofar as the numbers are more limited, as in a unit, sabotage could be a concern. In addition, cooperation, esprit de corps, and unit cohesion are important in the military, and pay differences can dampen or reduce overall team effectiveness, although evidence of this in the military context is lacking.[13]

Finally, other elements of compensation could affect the skewness of military compensation. After basic pay, the next largest element of cash compensation given to most members is BAH. While BAH varies by grade (as well as by dependents' status), location is the primary driver of the monthly BAH rate. BAH can dampen or accentuate the skewness of cash compensation depending on where members of different ranks tend to be located. Because BAH rates are not set with force-management considerations, such as performance incentives, in mind, the BAH-setting process can inadvertently and inefficiently alter the skewness of military compensation.

The discussion so far has focused on pay spreads across grades that are associated with promotion, but pay differences within a grade (e.g., intra-grade pay spreads or longevity increases) must also be determined. Changes in pay within a grade should also be structured to motivate effort and induce the proper sorting within the Army. Intra-grade pay should rise to some extent with experience to provide continuing skill acquisition and performance incentives (at least when coupled with minimum performance standards for retention). However, the longevity increases within a grade cannot be as large as the pay increases associated with promotion because individuals would be encouraged to "slum." And at some point, longevity

---

[13] As discussed in Lazear (2018), past studies found evidence that workers are less willing to be helpful to coworkers when promotion rewards are greater, and, consequently, he argued like Rosen (1992), there is a trade-off when increasing pay spreads between increasing individual effort incentives versus mitigating incentives for cooperative behavior.

increases should cease altogether so that those who have low promotion chances would be induced to leave voluntarily when it is in the services' interest that they do so. The report of seventh QRMC (Office of the Assistant Secretary of Defense (Force Management and Personnel, 1992) identified and recommended corrections of a number of inconsistencies between longevity and promotion-related increases in pay.

Finally, such personnel policies as up-or-out rules and minimum performance standards can play a positive role by (1) increasing effort and (2) inducing the voluntary departure of those who have low promotion chances. The extra turnover induced by up-or-out rules helps maintain promotion flows.

Talent Management and Entry Pay

The lack of lateral entry means that the military must access enough personnel at the entry level both to fill lower-level positions today and higher-level positions in the future. Because ability has an increasing effect on performance as individuals progress through the ranks, there should be a sufficient number of personnel in the entry cohort who have the capability to fill the upper-level positions and perform well in the future. But the military cannot just selectively recruit sufficient numbers of personnel who will successfully perform in the future because such ability is unobservable at entry. However, when entry pay is increased, the ability mix improves because higher entry pay attracts more applicants who have observable characteristics that are correlated with ability (education level and test scores), and the military can and does in fact screen for these characteristics. Pay beyond the entry level is also relevant, but because only about two-thirds of enlistees remain for only one enlistment, enlistment decisions will be based mostly on entry level and first-term pay.

The lack of lateral entry raises the required level of entry and first-term pay. If new entrants were required only to perform low-level tasks and were not needed to advance to the upper ranks, a less talented entry cohort could ably perform the lower-level tasks, which a lower-entry pay level would suffice to attract. Because of the requirement to raise entry pay to attract a more talented entrant cohort, many entrants are effectively overpaid because of the lack of lateral entry. (In the economist's jargon, they earn "economic rents," or payments in excess, of their next best alternatives.) How much higher entry pay should there be to ensure sufficient high-ability personnel to fill the upper ranks depends on the need for high ability in the upper ranks and the cost of paying economic rents to those fill lower-ranked jobs that do not require high ability (Asch and Warner, 1994).

Related to the higher level of entry pay is the implication that the military should proportionately access more entrants than would civilian employers that permit lateral entry so they can identify those with the talent to advance. This larger pool of entrants creates a "buffer stock" that enables the military to insure against talent shortfalls in the upper ranks.

Two other factors influence the level of entry pay. First is the positive or negative value that potential entrants place on nonmonetary factors associated with military life. Entry pay should be higher when the mean taste for military life is low among prospective entrants. The mean taste for service is, in turn, related to youths' perceptions of such factors as the prestige associated with military service, in-service living conditions, and the risk of death or injury. The recruiting experiences of the AVF period indicate that youths' perceptions of these factors have varied considerably over time and circumstance. As mentioned earlier, less than 10 percent of young people report a positive propensity to enlist (Joint Advertising and Marketing Research Studies, 2018).

The second factor is the degree of transferability of military-acquired job skills. Entry pay should be higher to provide an incentive to enlist when military job skills are less transferable. To some degree, bonuses serve the role of raising entry pay for those roles that have less-transferable skills. For example, the Army has historically paid sizable bonuses to enlistees in the combat arms occupations, which are nontransferable, but not to enlistees in electronics and maintenance, skills that are transferable. Higher bonuses may also be due to more-arduous conditions of service and the large requirements relative to supply of potential recruits.

Is the Current Basic Pay Table Skewed or Compressed?

This discussion directs us to the role of intra-rank and inter-rank pay spreads in providing incentives for performance and sorting high-ability personnel. How well does the current pay-table structure provide these incentives? Whether the current basic pay system is fully efficient in this regard is subject to debate. First, for the most part, intra-grade pay raises are not performance based but are provided in a lock-step fashion using longevity or tenure. Some use of performance-based intra-grade pay should be explored. Second, because longevity pay increases are based on time in service, early promotions provide only a temporary gain. A time-in-service–based system blunts the advantage to working harder and achieving an early promotion. A time-in-grade–based system of longevity increases has been recommended a number of times, including the tenth QRMC, and is currently being explored as part of the forthcoming 13th QRMC.

Whether the pay table is structured appropriately ultimately comes down to whether the Army and the other services are satisfied with the performance of military personnel. Even so, a direct look at the structure of the pay table can provide information on whether the pay table is skewed or appears compressed.

Figure 6 shows how basic pay varies with grade, assuming a hypothetical YOS history for enlisted personnel and officers.[14] For example, in 2018, pay of an E-1 is 56 percent of pay of an

---

[14] For enlisted, the profile for each year assumes average YOS at each grade as follows: E-1 at one YOS, E-2 at one YOS, E-3 at two YOS, E-4 at three YOS, E-5 at eight YOS, E-6 at 13 YOS, E-7 at 20 YOS, and E-9 at 25 YOS. For officers, the profile for each year assumes YOS at each grade as follows: O-1 at two YOS, O-2 at three YOS, O-3 at seven YOS, O-4 at 14 YOS, O-5 at 20 YOS, O-6 at 24 YOS, O-7 at 28 YOS, O-8 at 30 YOS, O-9 at 34 YOS, and O-10 at 36 YOS. These average YOS use data from the 1990s used in the Congressional Budget Office report

E-5, while the pay of an E-9 is 224 percent. The Career Compensation Act of 1949 (Public Law 81-351, 1949) established the general structure of the pay table, which remained relatively stable for enlisted personnel through 1958, except for the creation of grades E-8 and E-9 and for officers (except for the creation of grades O-9 and O-10). The structure of enlisted pay was skewed in 1958 in the sense that pay increases between E-1 and E-3 were smaller than later pay increases, making the structure relatively flat between E-1 and E-3. The pay increase rose at the promotion to E4 and was linear until E8. The pay increase rose again at the promotion to E9. In short, the 1958 pay structure resembled the example in Figure 5.

Junior entry pay for enlisted personnel was increased dramatically in 1971 in anticipation of the end of conscription. In 1958, E-1 pay was 40 percent of the pay of an E-5 but, after the 1971 targeted pay raise, it was 63 percent. The increase in enlisted entry pay at the start of the AVF was consistent with the guidance from the literature discussed above; higher entry pay was needed to draw a talented entry pool that could stay to fill higher-ranked positions later. The increase in entry pay tended to flatten or compress the overall 1972 enlisted pay structure, so that the difference between an E-1 and E-9 pay was less in 1972 than it was in 1958, but as mentioned, the need to increase entry pay would create this compression.

Between the start of the draft and 2018, Congress enacted a number of other targeted pay increases, focusing particularly on junior and mid-grade enlisted personnel and officers. Targeted pay raises were given in 1981 for enlisted personnel and for both enlisted and officers in 2000–2004 and again in 2007 when the 40-year pay table was created. These changes were in response to concerns about sustaining recruiting and retention, improving pay equity, and ensuring proper rewards to promotion, according to past QRMC reports such as the ninth QRMC (Office of the Under Secretary of Defense for Personnel and Readiness, 2002). The left panel of Figure 6 shows that the overall effect relative to the 1972 structure was to increase the convexity or skewness of the 2018 enlisted pay structure. The reward to achieving E-9 increased as did the reward for achieving E-4 to E-7, with some reduction in relative entry enlisted pay.

The officer pay structure is not convex or skewed, and pay compression has increased over time. During the draft era, as shown in Figure 6 for the year 1958, the officer pay structure was convex between the grades of O-1 and O-7 but became more compressed beginning with promotion to O-8. The convexity between O-1 and O-7 disappeared by the end of the draft era in 1972, and the structure beyond O-7 became even flatter over time. As of 2018, despite the periodic targeted pay raises during the draft era, the pay structure of senior officers is quite flat relative to more-junior officers. This observed structure reflects two key policies. First, beginning in 1981, senior officer pay in grades O-7 to O-10 was capped to executive-level pay for federal civilian workers. Second, no pay increases were given to those in paygrades O-7 and higher in 2015 and 2016. Capping senior officer pay reduced the skewness of the pay table,

---

Military Pay and the Rewards for Performance (Congressional Budget Office, 1995). Different YOS assumptions shift the curves up or down but do not change the shape appreciably.

thereby reducing incentives for performance along the entire officer hierarchy, not just for senior officers (Rosen, 1982). While comparisons of military and civilian executive compensation should be made with caution given their differing responsibilities and the legal factors affecting compensation (e.g., civilian executives can receive stock options), it is notable that studies of executive compensation and the compensation structures in large private corporations show that these structures are highly skewed, and executive compensation increases with size of firm (Rosen, 1981; Edmans, Gabaix, and Jenter, 2017).

Figure 6. Basic Pay Relative to Pay of an E-5 or O-3 (in Percentage), Selected Years

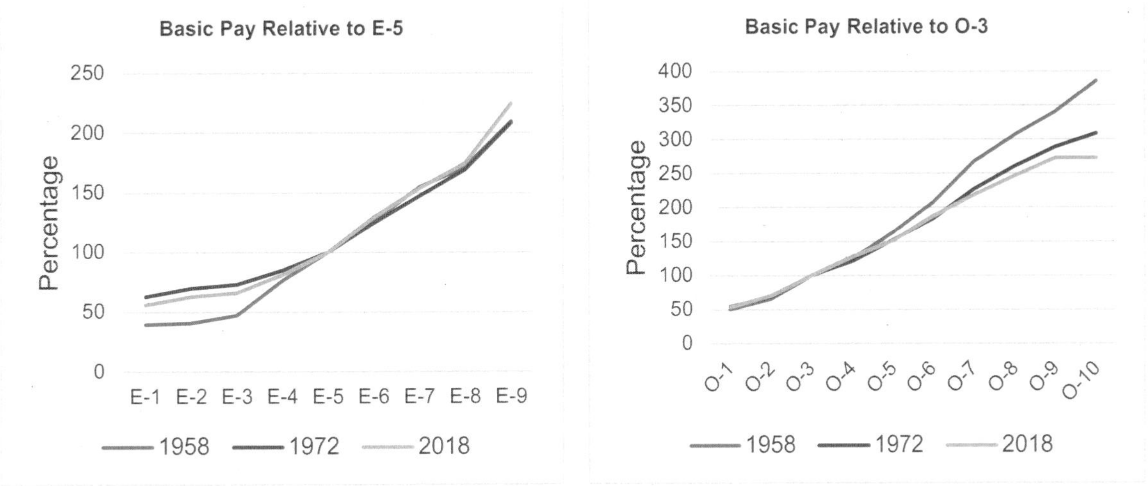

SOURCE: Congressional Budget Office, 1995; DoD, Office of the Under Secretary of Defense for Personnel and Readiness, Directorate of Compensation, 2000–2016.

In recent years, the services have become more concerned about talent management and ensuring flexibility of personnel policies to allow the services to recognize and reward high performers. Both Congress and the Army have introduced measures to increase the flexibility of military personnel management to better reward performance and meet emerging requirements in fields such as cyber. New flexibilities were introduced in the National Defense Authorization Act (NDAA) for FY 2019, and Congress has requested a review of the 1980 Defense Officer Personnel Management Act (DOPMA), which examines the current system and offers different reform proposals. At the same time, the Army's Talent Management Task Force is also considering ways to improve management flexibility. But any effort to improve retention, performance, and talent management should also consider the current structure of the basic pay table, especially for officers, because an appropriately structured table could complement efforts to address concerns about performance management.

The observed compressed officer pay structure suggests that factors that tend to reduce the required skewness has been particularly important to Congress and policymakers. For example, they may believe that officers give substantial value to the status and other nonmonetary factors associated with higher ranks or that senior officers already have a high taste for service, so that smaller additional monetary rewards are needed to motivate effort in the lower ranks, and less skewness is needed to motivate performance.

Area 3: Special and Incentive Pays

S&I pays are the primary way that compensation varies with occupation, assignment location, or type of duty for members in the same rank and YOS. The roughly 60 different S&I pays are intended to provide the services with flexibility to recognize persistently higher civilian pay for similar skills in the military, onerous or dangerous working conditions or assignments, specialized skills and proficiencies, temporary fluctuations in supply-and-demand conditions, and high training costs. The flexibility stems from the ability of the services to turn some of these pays on and off and to target them to specific groups of personnel to achieve potentially different experience or grade mixes across occupations. S&I pays are also considered to be a more cost-effective means of addressing recruiting and retention challenges than across-the-board increases in basic pay.

Research shows that bonuses targeted toward critical occupations are both effective and cost-effective. For example, Selective Reenlistment Bonuses (SRBs) are targeted to those who are eligible for reenlistment in critical specialties. Using data on first- and second-term retention in selected Army enlisted occupations between 2003 and 2007, Asch et al. (2010) found that a one-level increase in the SRB multiplier (which represents one month of basic pay per year of reenlistment) was estimated to increase the reenlistment rate by about 3 to 4 percentage points. Similarly, enlistment bonuses have been found to have a strong skill-channeling effect by inducing Army recruits to choose critical occupations. Studies also found that enlistment bonuses expand the recruiting market, estimating that a doubling of enlistment bonuses would increase number of high-quality Army enlistments by 4 to 13 percent (Knapp et al., 2018). Enlistment bonuses are found to be more cost-effective in expanding the market than across-the-board pay raises because the bonuses can be turned off and can be targeted to high-quality recruits (Asch et al., 2010) Analyses of special pays for officers, including mental health care providers and U.S. Air Force aviators, show that they increase retention and induce officers in these communities to select longer obligations (Hosek et al., 2017; Mattock et al., 2016).

S&I pays might be improved in three key areas. First, despite the large number of pays, S&I pays only account for $1 for every $20 in the 2019 defense military personnel budget. For individual members, S&I pays make up about 3 percent of total compensation and 5 percent of cash compensation, although the percentage is much higher in certain occupations (Defense Advisory Committee on Military Compensation, 2006). Put differently, the primary source of flexibility and efficiency in the compensation system turns out to be only a small fraction of cash compensation. This shortcoming is seen by the small differences in cash compensation across occupational areas for both Army enlisted personnel (Figure 7) and Army officers (Figure 8).

An important source of Army pay variation across members is attributable to bonuses and special pays (Asch, Hosek, and Martin, 2002), but pay variation is relatively small for members at similar points in their career. The relatively small share of S&I pay in compensation is likely attributable to a longstanding egalitarian philosophy that governs the setting military pay, specifically that everyone in the same grade and YOS should receive similar pay regardless of job type, location, and exposure to danger. But this egalitarian approach is expensive because significant cost savings can be realized from the pay targeting that S&I pays offer, specifically targeting pay to specific communities and circumstances rather than across the board. Past commissions have argued that S&I pays should comprise a larger share to improve pay flexibility and efficiency. So far, reallocating compensation from basic pay to S&I pays has been slow. One approach to accomplishing a reallocation could be to allocate a portion of the annual pay increase to S&I pays in a way that sustains retention across the force as well as in specific communities.

Figure 7. Army Enlisted Cash Compensation, by YOS and Occupational Area in 2004

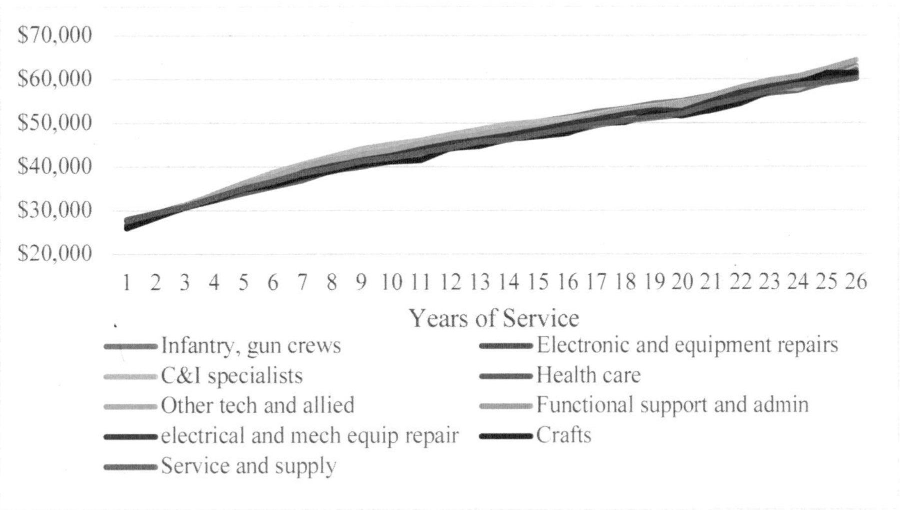

Figure 8. Army Officer Cash Compensation, by YOS and Occupational Area in 2004

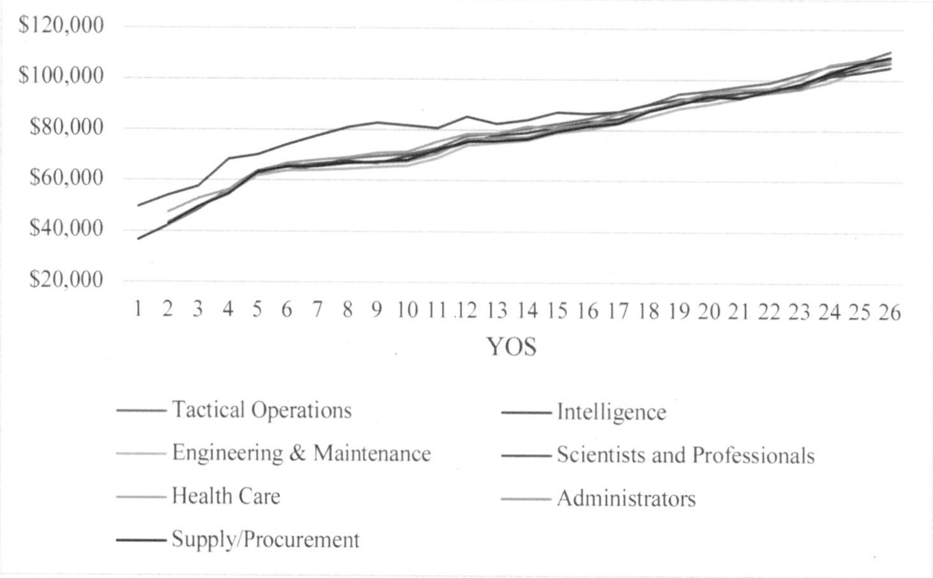

Second, S&I pays could be set to better support and reward performance. In a number of cases, these pays are flat monthly amounts fixed by law. For example, Special Duty Assignment Pay (SDAP) ranges from $75 to $450 per month. The amount varies with level of responsibility and demands of the assignment but not with performance. For instance, Army recruiters receive SDAP, but the amount is unrelated to the recruiter's productivity in achieving the recruiting mission. One approach for linking S&I pays to performance is to make these pay a function of pay grade and, therefore, promotion. Linking S&I pays to grade would reinforce the relationship between financial return and performance and promotion. SRBs are based on basic pay and therefore pay grade; they provide a higher benefit to those in higher grades at their reenlistment point. This helps tie SRBs to promotion and performance. That said, this relationship could be strengthened further by allowing any anniversary payments to increase if those who reenlist are promoted to a higher grade.[15] Another approach would be to explicitly base payment of some S&I pays on whether the soldier met performance criteria. For example, in the case of recruiting, different SDAP levels could be set so the payment depended on meeting recruiting goals over a period of time.

Third, S&I pays are more cost-effective when they include an incentive to select a longer obligation. For example, Hosek, Mattock, and Asch (2019) found that a dollar-for-dollar increase in aviation continuation pay (an S&I pay that depends on obligation length) has a larger effect on aviator retention than an increase in aviation continuation incentive pay (an S&I pay that is a flat

[15] Enlistment bonuses and reenlistment bonuses are sometimes paid in installments over the member's contract term length. The installment or anniversary payments are based on the pay, and therefore the pay grade, at the time of reenlistment. The text suggests allowing the anniversary payment to change as pay increases as a result of promotion.

monthly amount that does not depend on obligation length). Thus, to achieve a given change in retention, an S&I pay would have to increase more—and cost more—if it does not depend on a service obligation. The implication is that careful consideration should be given to identifying whether S&I pays that are currently monthly flat amounts could be made contingent on a service obligation.

Area 4: Deferred Compensation and Military Retirement

The creation of the BRS in the 2016 NDAA (and amended by the 2017 NDAA) represents the first major change to the military retirement system since the end of World War II. The legacy (pre-BRS) military retirement system dates back to 1947 and is known as a *defined-benefit plan*: The amount of the benefit is defined by a formula using basic pay, YOS, and a multiplier. As a result of modifications in 1981 and 1986, there are actually three legacy systems in effect. But the basic structure did not change; the system provided an immediate lifetime annuity to those who separate with 20 or more years of active duty but no benefits to those who separate with fewer than 20 years.[16] The BRS retains a defined-benefit plan from the legacy system and adds a defined-contribution plan and a new pay called *continuation pay* (CP). Under a defined-contribution plan, a percentage of earnings is paid regularly into an investment fund during the work-life of the future retiree and then paid out at retirement age as a lump sum or as an annuity.

BRS was introduced to modernize the legacy system and address several criticisms raised by past commissions and study groups. The next sections discusses how the BRS does this. It first provides a discussion of the broad objectives of any military retirement system and then discusses the criticisms of the legacy system in meeting these objectives, how the BRS addresses these criticisms, and potential issues in ensuring the success of the BRS.

Broad Objectives of the Military Retirement System

The purpose of military retired pay differs from that of retired pay in the civilian sector. Civilian firms are generally not bound by the military's lateral entry constraint, so they are not as concerned with generating turnover of older employees to create advancement opportunities for new hires. As a result, retired pay in the civilian sector is less of a tool to manage personnel flows and provide work-effort incentives and more of a vehicle to provide workers with tax-sheltered savings opportunities.

As discussed earlier, the lateral entry constraint means that the military should access and train large numbers of entrants before identifying for advancement those who have the talent to perform the higher-level tasks in the organization. Furthermore, the military should provide incentives for the most talented to stay and seek advancement and for others to leave after they

---

[16] The text describes the regular retirement system. The reserve retirement benefit generally begins at age 60.

discover that they are unsuitable for the upper-level positions. That is, it should provide the proper incentives for personnel to self-sort. Salop and Salop (1976) were the first to recognize the use of "two-part" compensation schemes as a self-selection device. One such "two-part" scheme is a system of (1) active pay and (2) deferred retirement benefits that are paid only to those who achieve a certain rank and longevity. Delayed vesting of immediate retired pay induces self-sorting because only those who think that they can achieve the requisite rank and longevity would stay early on while others would leave. Deferred retired pay can also motivate work effort, especially when combined with minimum performance standards for retention and up-or-out rules that prevent low-ranking personnel from staying long enough to collect retirement benefits.

The other major role for military retired pay is as a separation incentive. At some point, the military wants everyone, including the best personnel, to separate, even when they might still be individually very productive (i.e., their own productivity exceeds their pay). The longer individuals remain in the top positions, the slower the promotion rates would be for younger (and potentially equally able) personnel. Reduced promotion opportunities in the junior ranks would discourage work effort in those ranks, all else equal, and would cause those junior personnel with the best external opportunities (i.e., the more able) to leave. Without the proper inducement, senior personnel may not choose to leave voluntarily if their military pay exceeds their best private-sector alternatives, especially for those trained in the military-specific skills.

Retired pay can be used to induce voluntary separations of senior personnel to help control the age or experience structure of the force and to maintain promotion flows for younger personnel. As the 1948 Hook Commission stated:

> a sound retirement system is essential to solving the superannuation problem. The services must be kept young, vigorous, and efficient; a sound retirement plan with a proper compulsory retirement age will permit youth and brains to rise to the top in time to be effective. (Advisory Commission on Service Pay, 1948, p. 40)

The separations required to maintain personnel flows could be accomplished with other policy tools, notably more stringent up-or-out rules or with greater reliance on mandatory separations. But using these tools could be bad for morale, affecting recruiting, retention, and work effort, and they might require the payment of a "regret premium" to compensate for the prospect of involuntary separation. Furthermore, personnel faced with the prospect of involuntary separation are likely to engage in activities aimed at getting the policy relaxed (e.g., complaining to personnel managers, writing to representatives about the "unfairness" of the policy, engaging in lobbying activities). As a result of these "influence" activities, Congress might compel the services to modify their forces in unproductive or costly ways. By easing terminations from service with voluntary separations, retired pay helps weaken potential criticisms about the capriciousness or arbitrariness of mandatory separations and reduces these influence costs.

Failure to understand the separation function of retired pay is evident in the proposals of past commissions such as the Defense Business Board (2011), which recommended that the military system mimic the private sector—no benefits prior to age 60 except for those who complete a full 30-year career. Asch, Hosek, and Mattock (2014) show that retention would drop dramatically under such a system, necessitating a large active-duty pay raise to maintain the career force. But a large active-duty pay increase would necessitate heavier reliance on involuntary separation policy to control the experience distribution of the force.

The other purposes of retired pay are not unique. Motivating effort, improving retention in the mid-career, and inducing personnel to properly self-sort within the organization could be accomplished through an appropriately structured active-duty pay table and through other personnel policies.

Advantages of the Legacy Retirement System

The legacy system has several advantages. Perhaps the most important and obvious is that it has generally enabled the services to meet most of their manning requirements over the past 70 years despite tremendous changes in the military, technology and manpower requirements, DoD's wartime and peacetime posture, and the civilian population and economy. The system provides a predictable and stable career force to the services. By inducing mid-career personnel to stay in service—the so-called golden handcuffs of the retirement system—it facilitates high retention even when the external environment, such as business cycles, changes. And by inducing high turnover after 20 years of service, it helps the services maintain a young and vigorous force. Furthermore, it provides advantages to service members; the defined-benefit feature of the current system, backed by the federal government, is low risk and predictable, and the immediate annuity to qualified personnel provides a transition benefit to members as they embark on their second career in the civilian world. The system is also well understood and simple and rewards personnel with the same rank and experience in a similar way, thereby enhancing esprit de corps.

Criticism of the Legacy Retirement System

That said, the legacy system was controversial from the start and was the target of a number of critical analyses by DoD study groups over the years, including most recently the Military Compensation and Retirement Modernization Commission in 2015. Critics charged that the system is (1) excessively costly and unfair to taxpayers, (2) unfair to the vast majority of military entrants who do not serve long enough to receive retirement benefits, (3) inefficient, and (4) inflexible.

To the general public, the two most visible aspects of the legacy system are its cost and the relatively young ages of military retirees. The FY 2018 retired pay accrual for active-duty personnel was $18.1 billion for all of DoD and $5.8 billion for the Army (Office of the Under Secretary of Defense [Comptroller], 2018b; Department of the Army, 2018). The average retirement age of 45 for active-duty personnel in 2017 (DoD, Office of the Actuary, 2018) is far younger than the average retirement age of 64 for men and 62 for women among civilians (Munnell, 2015). This difference has led various study groups and commissions to argue that the legacy system induces those members with the most expertise to depart at young ages before the services would like for them leave and gives retirees "excessive" benefits at the expense of taxpayers.

Other critics charged that the legacy system was unfair to the 87 percent of military entrants who serve for fewer than 20 years. The military, in fact, is one of the few organizations exempted from the Employee Retirement Income Security Act (ERISA), the federal law that requires private-sector employers to vest employees in their retirement systems usually after five years of service. Some argued that the military should be brought under ERISA's early vesting requirements.

Implicit in the charge of "excessive cost" is the belief that military forces of the same quality could be obtained at less cost, meaning that the system is inefficient. Evidence suggests that members discount future dollars at a much higher rate than the government's borrowing rate. The estimated real personal discount factor ranges from 0.88 to 0.90 for enlisted personnel across the four services. That is, a dollar next year is worth 88 to 90 cents today. For officers, the estimates are similar across services—at 0.94 (see Asch, Hosek, and Mattock, 2014). In contrast, a typical government borrowing rate is around 3.5 percent, implying a discount factor of 0.965 is used (see, for example, Office of the Under Secretary of Defense [Comptroller], Chief Financial Officer, 2017b). Thus, members value a dollar in the future by less than the cost to the government of providing that dollar to military members. The implication is that the same force could be obtained at lower overall cost with more reliance on "up-front" (active-duty) pay and less reliance on retired pay. Taken to its extreme, the "up-front" view says that a retirement system is not necessary: The most efficient compensation system is an active pay–only system. Of course, some advocates of this line of reasoning recognize that it would be politically infeasible to eliminate the system altogether and therefore recommend a less generous system that conforms to ERISA guidelines for private-sector pension plans.

Finally, the system is alleged to hamper force management in several ways. First, it is believed to constrain the feasible experience distributions of the force. Under existing procedures, the services determine how many personnel they need in each paygrade using a variety of factors and then obtain desired experience distributions by translating requirements by paygrade into a YOS distribution. But the "desired" force structures are largely based on what can be supported with the retention patterns produced by the compensation system and not necessarily by the force structures the services would choose without the constraints imposed by the current compensation system. Force planning seems to follow a "tail wagging the dog" strategy, in which the compensation system determines the required force requirements instead of the other way around.

Furthermore, although one would expect the services to need quite different experience levels in different skill areas, the system has produced similar experience distributions across the broad spectrum of military skills. Some skills require "youth and vigor," while others do not. The Army and Marine Corps, for example, need large numbers of young personnel in the Combat Arms but relatively few at the more senior levels. Mid-level personnel in the combat skills who are not yet vested in the legacy retirement system are often assigned to other "infrastructure" jobs for the remainder of their 20-year careers. The Air Force and Navy, by contrast, require more-experienced personnel. Balancing very different demands for personnel is complicated by the current "one-shoe-fits-all" system.

How Does the BRS Address These Criticisms?

Notwithstanding such proposals as the 2011 Defense Business Board's that recommended replacing the legacy system with a defined contribution system, most recent commissions and study groups, including those that led to the BRS, have argued that the criticisms of the legacy system could be addressed by replacing it with a hybrid system that also included an increase in active-duty pay or other forms of current compensation. A hybrid system is one that includes both a defined benefit and defined contribution system.

A hybrid system that also increases current compensation helps keep the strengths of the military's legacy system while improving on its deficiencies. It keeps the defined benefit component of the legacy system, albeit less generous, and so helps keep its advantages such as predictability of future benefits. The defined contribution plan vests after three YOS, much earlier than the legacy system, thereby allowing members with a short career to leave with a portable retirement benefit. Consequently, the system is more equitable in the sense that more members become vested in the BRS than the legacy system.

Increasing current compensation while at the same time reducing the defined benefit plan reduces the backloaded nature of the military compensation system. Front-loading compensation increases the efficiency of the compensation system because members, who tend to be young and discount future dollars at a higher percentage than what it costs the government to provide, value a dollar of currency over a dollar of deferred compensation. Front-loading compensation means the same retention can be achieved at less cost. Thus, a hybrid system that includes higher current compensation produces cost savings. Furthermore, insofar as the current compensation takes the form of special and incentive pays that can be targeted to communities in which higher retention is desirable, a hybrid system that includes additional current compensation can also improve force-management flexibility.

The BRS is a hybrid system in which higher current compensation takes the form of CP and therefore offers the advantages of a hybrid system. By introducing matching Thrift Savings Plan (TSP) contributions that vest early, after two YOS, a significantly larger fraction of members, especially enlisted members, would be vested in the military retirement system, thereby addressing equity concerns about the legacy retirement system. DoD estimates that under the legacy system, 19 percent of active-duty service members separate with a retirement benefit; under the BRS, the figure is 85 percent (DoD, 2017). Furthermore, Congress gave discretion for setting CP to the services, and CP can be targeted to specific communities if the services choose to do so, thereby embedding additional flexibility into the military retirement system. The BRS also reduces the DB multiplier from 2.5 percent to 2.0 percent and created CP, paid as early as 8 YOS. In doing so, the BRS shifts compensation from the end of the career to earlier in the career. Greater front-loading of military compensation is more efficient and produces cost savings; the same force can be achieved at less cost, as discussed above in the context of the criticism about the excessive cost of the legacy system.

The BRS also provides members more choice about military retirement benefit. Members can choose the amount of their TSP contribution, whether to take CP and commit to a service obligation, and to take part of their annuity in the form of a lump sum (not mentioned above). Insofar as members have different preferences regarding savings, retirement, and the form their retirement annuity takes, the greater choice set has the potential to improve member satisfaction with their retirement benefit.

Analysis of the cost and retention effects of precursors of the BRS, such as the 2015 Military Compensation and Retirement Modernization Commission proposal and the 2013 Office of the Secretary of Defense working group proposal (see Asch, Hosek, and Mattock, 2014) show that, in the steady state (when all members are under the BRS), the same Army enlisted and officer force sizes and experience mixes could be achieved at less cost to the Army than the legacy system. That is, the lower accrual charge associated with the new and less-generous DB more than offsets the additional costs associated with TSP contributions and CP. However, in the short term, the Department of the Treasury outlays increase because the outlays associated with CP and TSP contributions begin immediately, but the lower outlays associated with smaller retirement annuities for retirees are not realized until the out-years. Long-run cost savings are achieved more quickly when opt-in rates are higher among eligible members who are grandfathered into the legacy system, because when opt-in rates are higher, more members are under the system with the less costly DB system.

Concerns About the BRS

Three issues have been raised concerning the BRS. Two relate to whether the Army and the other services can sustain retention under this new system. First, some have argued that vesting after two years of service in a portable retirement benefit—the TSP—gives early-career service members an incentive to leave earlier. Second, Congress mandated that CP be set at a minimum of 2.5 times monthly basic pay (e.g., a 2.5 multiplier), but concerns have been raised about whether the Army's and the other services' selection of the minimum CP multiplier for officers is too low for this group. The third issue that has been raised concerns whether members who are eligible have taken advantage of the BRS. Specifically, the concern is that grandfathered Army soldiers and officers who would probably be better off under the BRS have failed to opt into the BRS and have, instead, selected to remain under the legacy system.

TSP and the Incentive to Leave Service

Because the portability and early vesting of the TSP increases the value of leaving the military, the BRS does increase the incentive to leave. Furthermore, the reduction in the DB multiplier from 2.5 percent to 2.0 percent reduces the value of staying, thereby also increasing the incentive to leave. But for junior and early-career personnel, these incentives are relatively weak, because the TSP benefit cannot be claimed until age 59.5 at the earliest, and the DB is claimed after 20 YOS. Offsetting the increased incentive to leave is also a stronger incentive to stay. First, the addition of CP at between eight and 12 YOS, which is given to those who make at least a three-year service obligation, increases the incentive to stay both for junior personnel looking forward to the future and those who are under a CP service obligation. An individual who completes the service obligation after ten YOS would then be drawn to stay in service by the 20-year DB retirement plan. In addition, while early vesting of the TSP increases the value of leaving, the matching contributions that would be foregone by a member who leaves provide an incentive to stay.

Setting CP Appropriately

Research shows that the BRS can balance these incentives, and retention can be sustained for officers and for enlisted personnel relative to the legacy retirement system, but only if CP multipliers are set appropriately. This leads to the second concern, specifically that CP multipliers might not be set appropriately to sustain retention. Predicted retention under the BRS for Army enlisted personnel is sustained relative to retention under the legacy system when the CP multiplier is at the congressional minimum of 2.5 (Asch, Mattock, and Hosek, 2017). That is, the optimized CP multiplier that sustains Army enlisted retention under the BRS is 2.5, as shown in Figure 9 (left).[17] But the optimized multiplier that sustains Army officer retention is 11—or about a year's worth of basic pay. Thus, the minimum CP is too low for Army officers and is not predicted to sustain AC retention in the steady state (Figure 10).

To the extent that the Army has set the officer CP multiplier at the minimum of 2.5, the analysis predicts a reduced Army force size and experience mix for officers in the steady state. The steady state occurs in 40 years when all members are under the BRS. In the transition years, retention problems might not show up immediately or might be relatively muted; new officers who are automatically enrolled in the BRS would be under an initial service obligation, and officers who opted into the BRS would tend to be relatively junior and may not be influenced by the CP until they reach around eight YOS. But in the long-term, absent any other change, the

[17] Analysis also shows that reserve participation is also sustained in the steady state under the BRS for Army Guard and Reserve forces (Asch, Mattock, and Hosek, 2019).

BRS would lead to lower retention unless the CP is set higher than the minimum. The Army should carefully monitor officer retention and plan for fast action to raise the CP multiplier if actual retention results are too low.

Figure 9. Army Enlisted (Left) and Officer (Right) Personnel Under Legacy System Versus Blended Retirement System with Optimized CP Multipliers

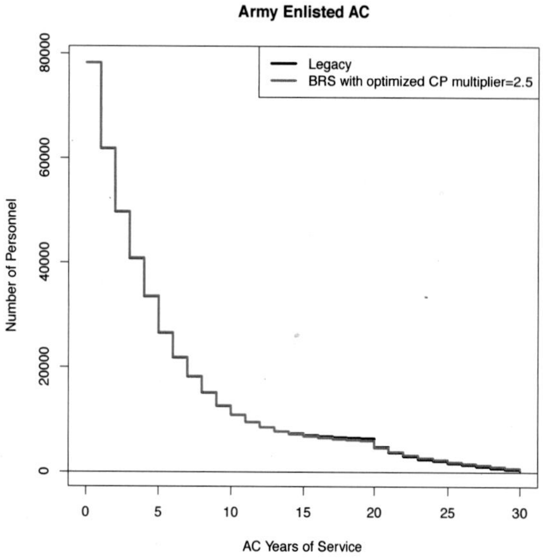

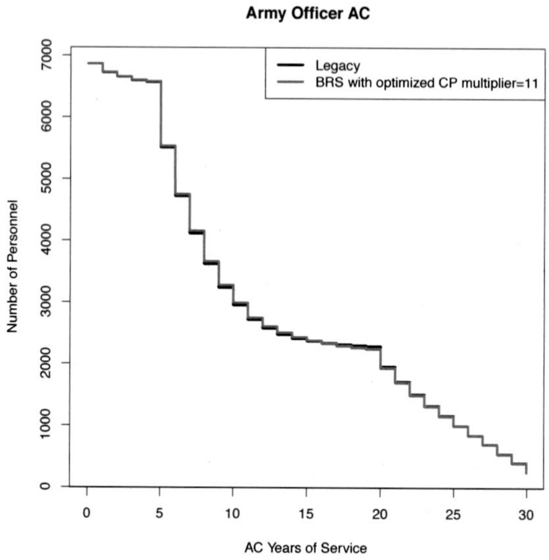

SOURCE: Asch, Mattock, and Hosek, 2017.

Figure 10. Army Officer Personnel Under Legacy System Versus Blended Retirement System with CP Multiplier Set at 2.5

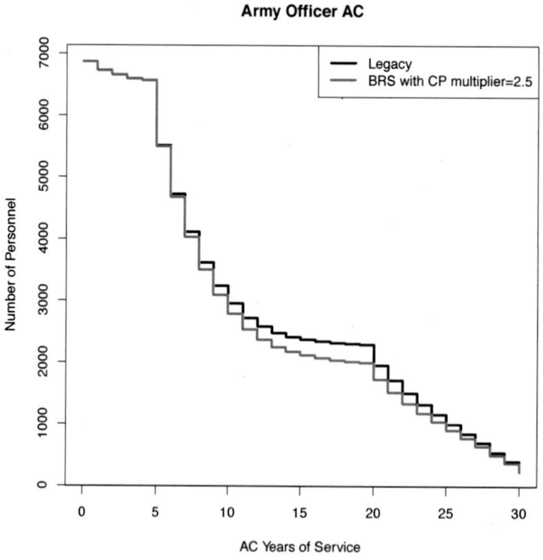

SOURCE: Asch, Mattock, and Hosek, 2017.

Thus, the adequacy of the minimum CP differs for officers and enlisted personnel. This difference is attributable to several factors. First, an individual in early or mid-career who looks forward to military retirement would expect a lower DB annuity under the BRS because of the reduced DB multiplier from 2.5 to 2.0 percent. Because officers have higher retention and a higher likelihood of reaching 20 YOS, they are more likely to experience the reduction in the DB multiplier than their enlisted counterparts. Second, officers earn higher basic pay and can expect a larger DB annuity if they become eligible at 20 YOS. Consequently, the reduction in the DB multiplier has a larger effect for officers than for enlisted personnel with a given number of YOS. Finally, we estimate a lower personal discount factor (around 0.88 to 0.90) for enlisted personnel than for officers (around 0.94). Thus, a reduction in future benefits, such as the reduction from the reduced DB multiplier under the BRS, has more of an effect on today's value for officers than for enlisted personnel. Both CP and TSP help to offset the negative effects of the reduced DB multiplier, but the optimized CP multiplier should be higher for officers than for enlisted personnel.

Low Opt-in Rates

All new accessions after January 1, 2018, are automatically enrolled into BRS. Members serving as of December 31, 2017, were grandfathered into the legacy system, with those with fewer than 12 YOS having the opportunity to switch to the BRS. The opt-in window ended December 31, 2018. Estimates through mid–December 2018 show that only 21.6 percent of active-duty Army members who were eligible to switch elected did (Philpott, 2019). The Army had the lowest switch rate of the four services (about one-half of the Marine Corps' 53.7 percent) and a switch rate below the rate predicted by RAND analysis (Asch, Mattock, and Hosek, 2017). Far fewer junior enlisted members have elected to switch than anticipated (Jowers, 2018).

The reason for the low opt-in is unclear. The RAND analysis assumed members fully understood the elements of the BRS and the financial implications of choosing the BRS versus the legacy system. It further assumed that members would make a rational choice using complete information that was provide, although the model recognized that members have imperfect information about the future.

It is possible that the financial education was not as effective as it could have been or that younger service members have an unrealistic (i.e., overly optimistic) assessment of their likelihood of reaching retirement. Because of the lower-than-expected opt-in, members did not act in their own best interest (i.e., they left money on the table). Because only about 10 percent of Army-enlisted entrants reach 20 YOS, these junior members would have been better off taking the BRS, since they would become eligible for the TSP, a portable benefit available after completing two years.

More information is needed to better understand why more-junior Army soldiers (as well as Navy sailors and Air Force airmen) did not switch to the BRS, while junior marines did. If it turns out that members made misinformed decisions and would value another opportunity to switch, DoD and Congress might consider opening up another opt-in window and implement appropriate actions to address reasons for the original lower-than-expected opt-in rates. Reopening the window would not be unprecedented. Eligible federal civil-service workers covered by the Civil Service Retirement System (CSRS) were given the opportunity to switch to the Federal Employees Retirement System (FERS), a plan that includes the TSP, when FERS was introduced in the 1980s. However, because of dramatic increases in stock market returns in the 1990s, many CSRS employees felt that they had "missed out," so Congress allowed a new opt-in period in the late 1990s. That said, the Office of the Secretary of Defense has taken a neutral stance about opt-in and had no official target for the degree of opt-in. So if a new window is reopened to give those who "missed out" another chance, the impetus would need to come from the services, such as the Army.

Emerging Compensation Issues

Congress is reviewing ways to increase flexibilities available to the services to manage their officer populations, focusing on reforms to personnel-management policies stipulated by the 1980 DOPMA. The 2018 NDAA asked for a DoD review of how the current system operates and for different reform proposals. The 2019 NDAA included DOPMA reforms that would give constructive credit to facilitate lateral entry, make permanent a career-intermission program that allows breaks in service, and other changes. Some observers have argued for a complete overhaul of DOPMA (Kane, 2018) while others, including LTG Thomas Seamands, have stated that DOPMA likely remains valid, but that it creates challenges in accommodating unique career paths, such as for those in cyber fields (Seamands, 2018). A review is currently underway, and RAND is assisting with that effort.

Any future authority to improve flexibility should also consider whether and how military compensation should change. The compensation system has performed remarkably over time, despite dramatic changes in the military and the environment in which it operates. Yet the desire for DOPMA reform raises questions about whether the current compensation structure is an obstacle to such reform. The forthcoming 13th QRMC is considering the advisability of replacing the current time-in-service pay table with a time-in-grade pay table. A time-in-grade table has been suggested as a way to better facilitate lateral entry (Chu, 2018) as well as to reward better performers (Defense Advisory Committee on Military Compensation, 2006; Office of the Secretary of Defense for Personnel and Readiness, 2008). However, appropriately structured constructive credit might be equally effective. More information is needed.

A second emerging issue is related to the efficiency of military compensation, specifically how the cost of military retirement liabilities to the services is computed. The retirement-accrual charge system was introduced in 1984 and was intended to bring visibility to the military-retirement liability resulting from current personnel policy decisions. The problem is that the current system produces inaccurate estimates of each service's accrual charge, including the

Army. This is because the accrual charge uses a combined, force-wide retention profile rather than service-specific profiles.

Hosek, Asch, and Mattock (2017) present empirical estimates of the inaccuracy in each service's total and marginal accrual charges under the current system compared with a system that sets service-specific accrual charges. This is done for the current military retirement benefit system and the reformed retirement benefit system recommended by the Military Compensation and Retirement Modernization Commission, a precursor of the BRS. The estimates show that the accrual charge is too high for three services and too low for one service. For example, the Army's accrual charge is estimated to be $380 million per year too high in 2015 dollars. Implementing service-specific accrual charges with separate charges for officers and enlisted personnel would eliminate these charge inaccuracies.

Section 1002 of S. 1519, a bill for the NDAA for FY 2018, proposed a service-specific military retirement accrual charge system, but the provision was not adopted (U.S. Congress, 2018). However, a conference report requested a U.S. Government Accountability Office (GAO) assessment of the effects of a service-specific accrual charge system. The GAO study, released December 2018, built on the 2017 Hosek, Asch, and Mattock report as well as a study conducted by DoD's Office of Cost Assessment and Program Evaluation (CAPE) described in the GAO report (GAO, 2018). CAPE also concluded that a service-specific accrual system would enforce better governance by providing more-accurate budget estimates and provide incentives to make better trade-offs in defense resource management. However, GAO reported that the Office of the Actuary stated that switching to a service-specific system would involve significant effort and possibly pose an operational risk associated with disaggregating funding requirements. Ultimately GAO concluded that the full effects of switching to a service-specific system was unknown. At this point, it is unclear how Congress will act in response to this information.

A third emerging issue is whether efficiency of the military compensation system would be improved by moving away from a basic pay and allowance system to a salary system that included basic pay, allowances, and the tax advantage. One of the arguments in favor of a salary system is that it could be more efficient. As mentioned, BAH, the largest element of cash compensation after basic pay for most members, is largely out of the control of defense managers and is a result of a housing survey. Rolling up BAH and other allowances into a salary system could allow the setting of military pay to more efficiently respond to force-management considerations, similar to the setting of pay in the civilian sector. Another argument is that it could improve the transparency of the military compensation system (Pleeter, 2001). A salary system could increase DoD costs depending on how the salary system accounts for the current distinction between those with and without dependents in setting BAH and the computation of the tax advantage. The forthcoming 13th QRMC was given the mandate to study the feasibility and advisability of a salary system for military personnel.

A fourth emerging issue is the potential of using compensation to a greater extent to induce more volunteerism within the military. The foundation of the AVF is the use of compensation and other policies to induce civilians to volunteer for service, and then once in service, to voluntarily remain in service. However, other aspects of service are anything but voluntary. For example, the matching of personnel to assignments or locations is often done through bureaucratic methods whereby personnel may end up in undesirable assignments or locations. In other words, there is a "share-the-pain policy": If someone takes a bad assignment today, he or she will get a better one tomorrow. This is a kind of "command-and-control" philosophy that is better suited for a draft military than an AVF.

A way that innovation in compensation policy could make the services and their personnel better off is through the use compensation to induce volunteers for onerous or dangerous assignments. The Navy piloted a test of this approach in its use of assignment incentive pay (AIP) in the early 2000s. In the pilot, Navy AIP rates were set in an auction using sailor willingness to go to particular locations and the number of unfilled billets (i.e., supply and demand). AIP was found to be more efficient than the share-the-pain approach. Specifically, Golfin, Lien, and Gregory (2004) and Golfin (2006) found that, when involuntary assignments fell to 5 percent from 100 percent, the Navy saved money (estimated at $2,200 per billet in 2006), and manning improved in sea-duty jobs. Furthermore, the auction system allowed member differences in their preferences and willingness to take assignments in different locations to be incorporated into the assignment process, thereby increasing the likelihood of member satisfaction with service and possibly their retention. In the future, more elements of compensation could be based on market mechanisms.

A final issue is improving the integration of the active and reserve compensation system. Because the reserve component has been used to a greater extent in an operational capacity, especially in Iraq and Afghanistan, policymakers have paid more attention to how reserve and active compensation compare and whether the compensation systems are perceived as equitable, while still recognizing that reserve service is part time during periods of nonactivation. One of the main areas of focus has been on differences and perceived inequities in the reserve versus active retirement system. Reserve retirement benefits under the legacy retirement system generally begin at age 60, whereas active benefits begin immediately upon separation with 20 or more years of service. Several studies, including the 11th QRMC (Office of the Under Secretary of Defense for Personnel and Readiness, 2012), have considered proposals that would reshape the reserve retirement system to make it more similar to the active system. For example, RAND Arroyo Center assessed an Army proposal that would allow the immediate availability of reserve retirement benefits upon separation with 20 or more years of service (Mattock et al., 2014). The report found that it would be possible to sustain the size of the Army active and reserve components with an overall cost savings of $800 million per year to the Army, in 2013 dollars, but that the experience mixes of both components would change. The 11th QRMC also considered proposals that would make the active and reserve retirement systems more similar, and again found that it could be possible to sustain retention at less cost (Office of the Under Secretary of Defense for Personnel and Readiness, 2012).

BRS is step toward making the active and reserve retirement systems more similar. Although it continues to be the case that reserve component members cannot begin DB benefits until age 60 in general, DB benefits are less generous under the BRS, and both active and reserve personnel receive TSP contributions and CP under the BRS.

Conclusion

A challenge to changing the military compensation system is that reforms must be done in coordination with the sister services, the Office of the Secretary of Defense, and Congress. The Army cannot act alone. But the Army can help provide the impetus for change. As discussed in this report, there are a number of areas that could be changed to improve member compensation satisfaction and to increase efficiency by enabling the Army and the other services to achieve retention and other readiness objectives at less cost and increase incentives for productivity and retention for the same cost.

The key recommendations and areas for consideration are as follows:

- Ascertain the Army's optimal mix of recruiting resources when relative military pay changes. The Army did not increase recruit aptitude as military and civilian pay rose in recent years, possibly because it kept recruiting quality goals constant as military pay increased, so recruiters were not induced to be more selective or perhaps the Army chose to reduce recruiting resources such as enlistment bonuses. Either way, the Army needs to identify why and when it should choose a different resource strategy and quality objective than the other services.
- Reevaluate the pay-adjustment mechanism. The ECI has clear disadvantages. Its functioning as an appropriate pay-adjustment mechanism and alternative approaches, including the DECI, should be reassessed with more-recent data.
- Consider increasing performance incentives embedded in the pay table. First, some use of performance-based intra-grade pay should be explored, including the possibility of a time-in-grade pay table. Second, the officer pay structure is compressed, and pay compression has increased over time, so inter-grade performance incentives for officers are dampened and occur through nonmonetary factors. Greater incentives associated with promotion should also be explored for officers, perhaps in conjunction with DOPMA reform.
- Improve the setting of S&I pay in the following ways:
  - To increase pay flexibility and efficiency, S&I pays should comprise a larger share of cash compensation. One approach would be to reallocate part of the basic pay increase toward higher S&I pay budgets.
  - Consider basing some S&I pays explicitly on soldiers meeting specific performance criteria and making S&I pays that are currently flat amounts per month contingent on a service obligation.
- Ensure the appropriate setting of CP for officers so that retention is sustained under the BRS and ascertain why Army opt-in rates were lower than the other services.
- Consider opening another opt-in window to permit members to choose the BRS.

- Increase the efficiency of the military retirement system by reforming the retirement accrual charge system to make it differ for officers and enlisted members and by service.
- Consider ways of using compensation to a greater extent to induce more volunteerism and greater efficiency of compensation within the Army, such as the unit or location assignment system.
- Recognize that changes to DOPMA to improve management flexibility should also consider whether and how military compensation should change.

Abbreviations

AC	active component
AFQT	Armed Forces Qualification Test
AIP	Assignment incentive pay
AVF	all-volunteer force
BAH	basic allowance for housing
BAS	basic allowance for subsistence
BRS	Blended Retirement System
CP	continuation pay
CSRS	Civil Service Retirement System
DB	defined benefit
DECI	Defense Employment Cost Index
DoD	U.S. Department of Defense
DOPMA	Defense Officer Personnel Management Act
DRM	Dynamic Retention Model
ECI	Employment Cost Index
ERISA	Employee Retirement Income Security Act
FERS	Federal Employees Retirement System
FY	fiscal year
GAO	U.S. General Accountability Office
MPA	Military Personnel Account
NDAA	National Defense Authorization Act
QRMC	*Quadrennial Review of Military Compensation*
RMC	regular military compensation
S&I	special and incentive
SDAP	Special Duty Assignment Pay
SRB	Selective Reenlistment Bonus

TSP	Thrift Savings Plan
YOS	year of service

References

Advisory Commission on Service Pay, *Career Compensation for the Uniformed Services: A Report and Recommendation for the Secretary of Defense*, Washington, D.C.: U.S. Government Printing Office, December 1948. As of February 1, 2019: https://babel.hathitrust.org/cgi/pt?id=uc1.b4244557;view=1up;seq=9

Asch, Beth J., Paul Heaton, James Hosek, Paco Martorell, Curtis Simon, and John T. Warner, *Cash Incentives and Military Enlistment, Attrition, and Reenlistment*, Santa Monica, Calif.: RAND Corporation, MG-950-OSD, 2010. As of January 31, 2019: https://www.rand.org/pubs/monographs/MG950.html

Asch, Beth J., James Hosek, and Craig Martin, *A Look at Cash Compensation for Active-Duty Military Personnel*, Santa Monica, Calif.: RAND Corporation, MR-1492-OSD, 2002. As of February 1, 2019: https://www.rand.org/pubs/monograph_reports/MR1492.html

Asch, Beth J., James Hosek, and Michael G. Mattock, *Toward Meaningful Military Compensation Reform: Research in Support of DoD's Review*, Santa Monica, Calif.: RAND Corporation, RR-501-OSD, 2014. As of January 31, 2019: https://www.rand.org/pubs/research_reports/RR501.html

Asch, Beth J., James Hosek, Michael G. Mattock, and Christina Panis, *Assessing Compensation Reform: Research in Support of the 10th Quadrennial Review of Military Compensation*, Santa Monica, Calif.: RAND Corporation, MG-764-OSD, 2008. As of January 31, 2019: https://www.rand.org/pubs/monographs/MG764.html

Asch, Beth J., Michael G. Mattock, and James Hosek, *The Blended Retirement System: Retention Effects and Continuation Pay Cost Estimates for the Armed Services*, Santa Monica, Calif.: RAND Corporation, RR-1887-OSD/USCG, 2017. As of January 31, 2019: https://www.rand.org/pubs/research_reports/RR1887.html

Asch, Beth J., Michael G. Mattock, and James Hosek, *Effects of the Blended Retirement System on United States Army Reserve Participation and Cost*, Santa Monica, Calif.: RAND Corporation, RR-2591-A, 2019. As of August 28, 2019: https://www.rand.org/pubs/research_reports/RR2591.html

Asch, Beth J., John A. Romley, and Mark E. Totten, *The Quality of Personnel in the Enlisted Ranks*, Santa Monica, Calif.: RAND Corporation, MG-324-OSD, 2005. As of October 2, 2018: https://www.rand.org/pubs/monographs/MG324.html

Asch, Beth J., and John T. Warner, "A Theory of Compensation and Personnel Policy in Hierarchical Organizations with Application to the United States Military," *Journal of Labor Economics*, Vol. 19, No. 3, July 2001, pp. 523–562.

Asch, Beth J., and John T. Warner, *A Theory of Military Compensation and Personnel Policy*, Santa Monica, Calif.: RAND Corporation, MR-439-OSD, 1994. As of August 28, 2019: https://www.rand.org/pubs/monograph_reports/MR439.html

Chu, David, "Reconsidering the Defense Officer Personnel Management Act," testimony to the Subcommittee on Personnel, Committee on Armed Services, Washington, D.C.: U.S. Senate, January 22, 2018. As of August 28, 2019: https://www.armed-services.senate.gov/imo/media/doc/Chu_01-24-18.pdf

Congressional Budget Office, *Military Pay and the Rewards for Performance*, Washington, D.C., December 1995. As of August 28, 2019: https://www.cbo.gov/sites/default/files/104th-congress-1995-1996/reports/doc30-entire.pdf

Defense Advisory Committee on Military Compensation, *The Military Compensation System: Completing the Transition to an All-Volunteer Force*, Arlington, Va., April 28, 2006. As of January 31, 2019: http://www.airforcemag.com/DocumentFile/Documents/2006/dacmc042806finalreport.pdf

Defense Business Board, *Report to the Secretary of Defense: Modernizing the Military Retirement System, Report FY11–05*, Washington, D.C., July 21, 2011. As of January 31, 2019: https://dbb.defense.gov/Portals/35/Documents/Reports/2011/FY11-5_Modernizing_The_Military_Retirement_System_2011-7.pdf

Defense Health Agency, *Evaluation of the TRICARE Program: Fiscal Year 2018 Report to Congress, Access, Cost, and Quality Data Through Fiscal Year 2017*, Falls Church, Va., May 9, 2018. As of May 9, 2019: https://www.health.mil/Military-Health-Topics/Access-Cost-Quality-and-Safety/Health-Care-Program-Evaluation/Annual-Evaluation-of-the-TRICARE-Program

Defense Manpower Data Center, "Armed Forces Strength Figures for November 30, 2018," webpage, November 30, 2018.

Department of the Army, U.S. Department of Defense, *Fiscal Year (FY) 2019 Budget Estimates: Military Personnel, Army Justification Book*, Washington, D.C.: U.S. Department of Defense, February 2018. As of January 31, 2019: https://www.asafm.army.mil/Portals/72/Documents/BudgetMaterial/2019/Base%20Budget/Military%20Personnel/Military%20Personnel%20Army.pdf

DoD—*See* U.S. Department of Defense.

Edmans, Alex, Xavier Gabaix, and Dirk Jenter, *Executive Compensation: A Survey of Theory and Evidence*, Cambridge, Mass.: National Bureau of Economic Research, Working Paper 23596, July 2017. As of September 11, 2019: https://www.nber.org/papers/w23596.pdf

Ford, M., B. Griepentrog, K. Helland, and S. Marsh, *The Strength and Variability of the Military Propensity-Enlistment Relationship: Evidence from 1995–2003*, Washington, D.C.: Joint Advertising, Market Research, and Studies, Office of the Undersecretary of Defense for Personnel and Readiness, U.S. Department of Defense, JAMRS Report No. 2009-005, April 2009.

GAO—*See* U.S. Government Accountability Office.

Golfin, Peggy A., *Manning Under AIP*, Alexandria, Va.: Center for Naval Analyses, CAB D0014440.A1/Final, June 2006. As of January 31, 2019: https://www.cna.org/CNA_files/PDF/d0014440.a1.pdf

Golfin, Peggy A., Diana S. Lien, and David Gregory, *Evaluation of the Assignment Incentive Pay (AIP) System*, Alexandria, Va.: Center for Naval Analyses, CAB D0010240.A2/Final, June 2004. As of January 31, 2019: https://www.cna.org/CNA_files/PDF/D0010240.A2.pdf

Gotz, Glenn A., and John McCall, *A Dynamic Retention Model for Air Force Officers: Theory and Estimates*, Santa Monica, Calif.: RAND Corporation, R-3028-AF, 1984. As of January 31, 2019: https://www.rand.org/pubs/reports/R3028.html

Hogan, Paul F., and Rita Furst Seifert, "Marriage and the Military: Evidence That Those Who Serve Marry Earlier and Divorce Earlier," *Armed Forces and Society*, Vol. 36, No. 3, April 2010, pp. 420–438.

Hosek, James, Beth J. Asch, and Michael G. Mattock, *Toward Efficient Military Retirement Accrual Charges*, Santa Monica, Calif.: RAND Corporation, RR-1373-A, 2017. As of January 31, 2019: https://www.rand.org/pubs/research_reports/RR1373.html

Hosek, James, Beth J. Asch, Michael G. Mattock, and Troy D. Smith, *Military and Civilian Pay Levels, Trends, and Recruit Quality*, Santa Monica, Calif.: RAND Corporation, RR-2396-OSD, 2018. As of February 1, 2019: https://www.rand.org/pubs/research_reports/RR2396.html

Hosek, James, and Paco Martorell, *How Have Deployments During the War on Terrorism Affected Reenlistment?* Santa Monica, Calif.: RAND Corporation, MG-873-OSD, 2009. As of January 31, 2019: https://www.rand.org/pubs/monographs/MG873.html

Hosek, James, Michael G. Mattock, and Beth J. Asch, *A Wage Differential Approach to Managing Special and Incentive Pay*, Santa Monica, Calif.: RAND Corporation, RR-2101-OSD, 2019. As of February 1, 2019: https://www.rand.org/pubs/research_reports/RR2101.html

Hosek, James, Shanthi Nataraj, Michael G. Mattock, and Beth J. Asch, *The Role of Special and Incentive Pays in Retaining Military Mental Health Care Providers*, Santa Monica, Calif.: RAND Corporation, RR-1425-OSD, 2017. As of February 1, 2019: https://www.rand.org/pubs/research_reports/RR1425.html

Hosek, James, Christine E. Petersen, Jeannette Van Winkle, and Hui Wang, *A Civilian Wage Index for Defense Manpower*, Santa Monica, Calif.: RAND Corporation, R-4190-FMP, 1992. As of January 31, 2019: https://www.rand.org/pubs/reports/R4190.html

Hosek, James, and Shelley MacDermid Wadsworth, "Economic Conditions of Military Families," *The Future of Children*, Vol. 23, No. 2, Fall 2013. As of April 3, 2019: https://files.eric.ed.gov/fulltext/EJ1018366.pdf

Joint Advertising and Marketing Research Studies, *Summer 2017 Propensity Update: Youth Poll Study Findings*, Washington, D.C.: Office of the Undersecretary of Defense for Personnel and Readiness, U.S. Department of Defense, February 2018, not available to the general public.

Jowers, Karen, "Not Many Troops Are Opting into the New Retirement System," *Military Times*, October 22, 2018. As of January 31, 2019: https://www.militarytimes.com/pay-benefits/2018/10/22/not-many-troops-are-opting-into-the-new-retirement-system/

Kane, Tim, "Statement of Dr. Tim Kane Before the Senate Armed Services Committee Hearing on Officer Personnel Management and the Defense Officer Personnel Management Act of 1980," hearing before the Armed Services Committee, Washington, D.C.: U.S. Senate, January 24, 2018. As of August 29, 2019: https://www.armed-services.senate.gov/imo/media/doc/Kane_01-24-18.pdf

Kapp, Lawrence, *Defense Primer: Military Pay Raise*, Washington, D.C.: Congressional Research Service, IF10260, August 1, 2018. As of August 29, 2019: https://www.hsdl.org/?abstract&did=814440

Knapp, David, Bruce R. Orvis, Christopher E. Maerzluft, and Tiffany Tsai, *Resources Required to Meet the U.S. Army's Enlisted Recruiting Requirements Under Alternative Recruiting Goals, Conditions, and Eligibility Policies*, Santa Monica, Calif.: RAND Corporation, RR-2364-A, 2018. As of February 1, 2019: https://www.rand.org/pubs/research_reports/RR2364.html

Lazear, Edward P., "Compensation and Incentives in the Workplace," *Journal of Economic Perspectives*, Vol. 32, No. 3, Summer 2018, pp. 195–214.

Lazear, Edward P., and Sherwin Rosen, "Rank Order Tournaments as Optimal Labor Contracts," *Journal of Political Economy*, Vol. 89, No. 5, October 1981, pp. 841–864.

Lewin Group, Inc., *Qualified Military Available (QMA): Final Technical Report*, Falls Church, Va.: Lewin Group, Inc., October 10, 2013.

Mattock, Michael G., Beth J. Asch, and James Hosek, *Making the Reserve Retirement System Similar to the Active System: Retention and Cost Estimates*, Santa Monica, Calif.: RAND Corporation, RR-530-A, 2014. As of January 31, 2019: https://www.rand.org/pubs/research_reports/RR530.html

Mattock, Michael G., Beth J. Asch, James Hosek, Christopher Whaley, and Christina Panis, *Toward Improved Management of Officer Retention: A New Capability for Assessing Policy Options*, Santa Monica, Calif.: RAND Corporation, RR-764-OSD, 2014. As of January 31, 2019: https://www.rand.org/pubs/research_reports/RR764.html

Mattock, Michael G., James Hosek, Beth J. Asch, and Rita Karam, *Retaining U.S. Air Force Pilots When the Civilian Demand for Pilots Is Growing*, Santa Monica, Calif.: RAND Corporation, RR-1455-AF, 2016. As of February 1, 2019: https://www.rand.org/pubs/research_reports/RR1455.html

Munnell, Alicia H., *The Average Retirement Age—An Update*, Boston, Mass.: Center for Retirement Research at Boston College, No. 15-4, March 2015. As of January 31, 2019: http://crr.bc.edu/wp-content/uploads/2015/03/IB_15-4_508_rev.pdf

Office of Personnel Management, "Data, Analysis & Documentation: Federal Employment Reports: Full-Time Permanent Age Distributions," webpage, September 2017. As of January 31, 2019: https://www.opm.gov/policy-data-oversight/data-analysis-documentation/federal-employment-reports/reports-publications/full-time-permanent-age-distributions/

Office of the Assistant Secretary of Defense (Force Management and Personnel), *Report of the Seventh Quadrennial Review of Military Compensation*, Washington, D.C.: U.S. Department of Defense, August 21,1992.

Office of the Under Secretary of Defense (Comptroller), Chief Financial Officer, *Defense Budget Overview: United States Department of Defense Fiscal Year 2018 Budget Request*, Washington, D.C.: U.S. Department of Defense, May 2017a. As of January 30, 2019: https://comptroller.defense.gov/Portals/45/Documents/defbudget/fy2018/fy2018_Budget_Request_Overview_Book.pdf

Office of the Under Secretary of Defense (Comptroller), Chief Financial Officer, Fiscal Year 2017: *Military Retirement Fund Audited Financial Report*, Washington, D.C.: U.S. Department of Defense, November 6, 2017b. As of January 31, 2019: https://comptroller.defense.gov/Portals/45/documents/cfs/fy2017/13_Military_Retirement_Fund/FY2017_MRF_AFR_Final.pdf

Office of the Under Secretary of Defense (Comptroller), Chief Financial Officer, *Military Personnel Programs (M-1): Department of Defense Budget—Fiscal Year 2019*, Washington, D.C.: U.S. Department of Defense, February, 2018a. As of January 31, 2019: https://comptroller.defense.gov/Portals/45/Documents/defbudget/fy2019/fy2019_m1.pdf

Office of the Under Secretary of Defense (Comptroller), *National Defense Budget Estimates for FY 2019*, Washington, D.C.: U.S. Department of Defense, April 2018b. As of January 31, 2019: https://comptroller.defense.gov/Portals/45/Documents/defbudget/fy2019/FY19_Green_Book.pdf

Office of the Under Secretary of Defense for Personnel and Readiness, *Report of the Fifth Quadrennial Review of Military Compensation*, Washington, D.C.: U.S. Department of Defense, January 1984.

Office of the Under Secretary of Defense for Personnel and Readiness, *Population Representation in the Military Services*, Washington, D.C.: U.S. Department of Defense, fiscal years 1997–2016. As of January 31, 2019: https://www.cna.org/research/pop-rep

Office of the Under Secretary of Defense for Personnel and Readiness, *Report of the Ninth Quadrennial Review of Military Compensation*, Vols. I–V, Washington, D.C.: U.S. Department of Defense, March 2002. As of January 31, 2019: https://militarypay.defense.gov/Portals/3/Documents/Reports/9th_QRMC_Report_Volumes_I_-_V.pdf

Office of the Under Secretary of Defense for Personnel and Readiness, *Report of the Tenth Quadrennial Review of Military Compensation*, Vol. I: Cash Compensation, Washington, D.C.: U.S. Department of Defense, February 2008. As of January 31, 2019: https://militarypay.defense.gov/Portals/3/Documents/Reports/10th_QRMC_2008_Vol_I_Cash_Compensation.pdf

Office of the Under Secretary of Defense for Personnel and Readiness, *Military Compensation Background Papers: Compensation Elements and Related Manpower Cost Items, Their Purposes and Legislative Backgrounds*, 7th ed., Washington, D.C.: U.S. Department of Defense, November 2011. As of January 31, 2019: https://www.loc.gov/rr/frd/pdf-files/Military_Comp-2011.pdf

Office of the Under Secretary of Defense for Personnel and Readiness, *Report of the Eleventh Quadrennial Review of Military Compensation: Main Report*, Washington, D.C.: U.S. Department of Defense, June 2012. As of January 31, 2019: https://militarypay.defense.gov/Portals/3/Documents/Reports/11th_QRMC_Main_Report_FINAL.pdf?ver=2016-11-06-160559-590

Office of the Under Secretary of Defense for Personnel and Readiness, *Military Compensation Background Papers: Compensation Elements and Related Manpower Cost Items, Their Purposes and Legislative Backgrounds*, 8th ed., Washington, D.C.: U.S. Department of Defense, July 2018. As of January 31, 2019: https://militarypay.defense.gov/Portals/3/Documents/Reports/Mil-Comp_8thEdition.pdf?ver=2018-09-01-181142-307

Orvis, Bruce R., Michael Childress, and J. Michael Polich, *Effect of Personnel Quality on the Performance of Patriot Air Defense System Operators*, Santa Monica, Calif.: RAND Corporation, R-3901-A, 1992. As of February 1, 2019: https://www.rand.org/pubs/reports/R3901.html

Philpott, Tom, "'Blended' Military Retirement Plan Not as Popular as Projected," *Military.com*, January 3, 2019. As of January 31, 2019: https://www.military.com/daily-news/2019/01/03/blended-military-retirement-plan-not-popular-projected.html

Pleeter, Saul, "Cost of Living, Allowances and a Salary System," unpublished paper presented at the Western Economics Association Meetings, July 2001.

"Project A: The U.S. Army Selection and Classification Project—Special Issue," *Personnel Psychology*, Vol. 43, No. 2, June 1990. As of February 1, 2019: https://onlinelibrary.wiley.com/toc/17446570/1990/43/2

Public Law 81-351, Career Compensation Act of 1949, October 12, 1949.

Rosen, Sherwin, "The Economics of Superstars," *American Economic Review,* Vol. 71, No. 5, December 1981, pp. 845–858.

Rosen, Sherwin, "Authority, Control, and the Distribution of Earnings," *Bell Journal of Economics*, Vol. 13, No. 2, Autumn 1982, pp. 311–323.

Rosen, Sherwin, "The Military as an Internal Labor Market: Some Allocation, Productivity, and Incentive Problems," *Social Science Quarterly*, Vol. 73, No. 2, June 1992, pp. 227–237.

Salop, Joanne, and Steven Salop, "Self-Selection and Turnover in the Labor Market," *The Quarterly Journal of Economics*, Vol. 90, No. 4, November 1976, pp. 619–627.

Scribner, Barry L., D. Alton Smith, Robert H. Baldwin, and Robert L. Phillips, "Are Smart Tankers Better? AFQT and Military Productivity," *Armed Forces and Society*, Vol. 12, No. 2, Winter 1986, pp. 193–206.

Seamands, Thomas C., "Record Version Statement by LTG Thomas C. Seamands, Deputy Chief of Staff, G-1, United States Army Before the Personnel Subcommittee Senate Armed Services Committee on the Defense Officer Personnel Management Act," Washington, D.C.: U.S. Senate, 2nd. Sess., 115th Congress, January 24, 2018. As of August 29, 2019: https://www.armed-services.senate.gov/imo/media/doc/Seamands_01-24-18.pdf

U.S. Code, Title 37, Section 1009, Adjustments of Monthly Basic Pay, November 24, 2003.

U.S. Congress, 115th Cong., National Defense Authorization Act for Fiscal Year 2018, S. 1519, March 22, 2018.

U.S. Congress, Congressional Budget Office, *Evaluating Military Compensation*, Washington, D.C.: U.S. Government Printing Office, June 2007. As of January 31, 2019: https://www.cbo.gov/sites/default/files/110th-congress-2007-2008/reports/06-29-compensation.pdf

U.S. Department of Defense, *Frequently Asked Questions Regarding the New Blended Retirement System*, Washington, D.C., May 1, 2017. As of January 31, 2019: https://militarypay.defense.gov/Portals/3/Documents/BlendedRetirementDocuments/BRS%20Frequently%20Asked%20Questions%205.01.2017.pdf

U.S. Department of Defense, Office of the Actuary, *Valuation of the Military Retirement System: September 30, 2016*, Washington, D.C.: U.S. Department of Defense, June 2018. As of February 1, 2019: https://media.defense.gov/2018/Jun/08/2001928794/-1/-1/0/MRF%20VALRPT%202016%20[JUNE%202018].PDF

U.S. Department of Defense, Office of the Deputy Assistant Secretary of Defense for Military Community and Family Policy, *2016 Demographics: Profile of the Military Community*, Washington, D.C.: U.S. Department of Defense, 2016. As of January 31, 2019: https://download.militaryonesource.mil/12038/MOS/Reports/2016-Demographics-Report.pdf

U.S. Department of Defense, Office of the Under Secretary of Defense for Personnel and Readiness, Directorate of Compensation, *Selected Military Compensation Tables*, Washington, D.C., 2000–2016. Pay tables for years 2000 through 2016 available online. As of November 4, 2019: https://www.dfas.mil/militarymembers/payentitlements/Pay-Tables/PayTableArchives.html

U.S. Department of Labor, Bureau of Labor Statistics, "Employment Cost Trends," webpage, undated. As of September 11, 2019: https://www.bls.gov/ncs/ect/

U.S. Government Accountability Office, *Military Retirement: Service Contributions Do Not Reflect Service Specific Estimated Costs and Full Effect of Proposed Legislation Is Unknown*, Washington, D.C., GAO-19-195R, December 4, 2018. As of January 31, 2019: https://www.gao.gov/assets/700/695789.pdf

Vessey, John, "Thirty Years of an All-Volunteer Force: Personal Observations," in Barbara Bicksler, Curtis Gilroy, and John Warner, eds., *The All-Volunteer Force: Thirty Years of Service*, Washington, D.C.: Brassey's, 2004, pp. 335–342.

Warner, John T., *Thinking About Military Retirement*, Alexandria, Va.: Center for Naval Analyses, CRM D0013583.A1/Final, January 2006.

Warner, John T., and Beth J. Asch, "Chapter 13: The Economics of Military Manpower," in Sandler, Todd and Keith Hartley, eds., *Handbook of Defense Economics*, Vol. I, Amsterdam, The Netherlands: Elsevier, 1995, pp. 347–398.

Winkler, John D., Judith C. Fernandez, and J. Michael Polich, *Effect of Aptitude on the Performance of Army Communications Operators*, Santa Monica, Calif.: RAND Corporation, R-4143-A, 1992. As of October 2, 2018: https://www.rand.org/pubs/reports/R4143.html